Sensing of Non-Volatile Memory Demystified

Swaroop Ghosh
Editor

Sensing of Non-Volatile Memory Demystified

 Springer

Editor
Swaroop Ghosh
School of Electrical Engineering and
   Computer Science
Pennsylvania State University
State College, PA, USA

ISBN 978-3-030-07339-8   ISBN 978-3-319-97347-0   (eBook)
https://doi.org/10.1007/978-3-319-97347-0

This Springer imprint is published by the registered company Springer Nature Switzerland AG
The registered company address is: Gewerbestrasse 11, 6330 Cham, Switzerland

Preface

Research, development, and commercialization of emerging non-volatile memories (NVMs) are being aggressively pursued by the design community to supplement and/or substitute conventional volatile and NV memory technologies that are facing stiff scalability challenges. The emerging NVMs such as Magnetic RAM (MRAM), Spin-Transfer Torque RAM (STTRAM), Resistive RAM (RRAM) and Phase Change Memory (PCM) are already available in market as discreet chips. It has been forecasted by Yole that the emerging NVM market will continue to grow and find several niche applications in health care, banking, and day-to-day computing. One of the key aspects of the functioning of these NVMs is sensing. Since the emerging NVMs possess better reliability and faster access latency, they are positioned to replace embedded higher-level memory such as cache, main memory, and solid-state drives (SSDs). The well-known flash sensing techniques which are latency-intensive are not suitable for emerging NVMs. The sensing techniques for conventional charge-based memories such as SRAM and DRAM cannot be extended to emerging NVMs since majority of these memories are resistive in nature. The emerging NVMs face stiff sensing challenges due to process, voltage and temperature variations, and low operating voltage requirements. The desire to achieve high memory density places further restrictions on sense margin. Many conflicting design trade-offs exist to maximize the sense margin at the cost of area, power, and delay. A focused treatment on the sensing challenges associated with emerging NVMs is required to accelerate their research and development efforts.

This is a first book which covers the sensing of emerging NVMs spanning spintronic, resistive, phase-change, and ferroelectric memory technologies. Challenges such as shrinking margin between resistive states due to intrinsic and extrinsic sources of variations, noise and need for large memory capacity at tighter footprint, lower operating voltages, and faster access latency are discussed in detail in the context of sense amplifier design. Various state-of-the-art resilience enhancement techniques are also presented. The book also investigates the impact of application that poses new challenges and offers new opportunities for reliable sensing.

Why a New Book

Reliable sensing is a key to enable wide range of applications of emerging NVMs. The choice of memory technology determines the type of sensing technique (current-based or voltage-based) which is drastically different from their conventional memory counterparts. A multitude of interesting techniques are explored to maximize the sense margin in the face of vanishing gap between resistance states due to issues such as process variations. The students, industrial/academic researchers, and practicing engineers are largely ignorant about these new circuit-level innovations. There is lack of a book that specifically addresses the sensing challenges of emerging NVMs. Considering the importance of the topic, the book compiles outstanding contributions from prominent personalities in this area covering different aspects of memory-specific sensing and variation-tolerant design.

Unique Features

- Introduction to emerging NVMs, their operating mechanisms, and features.
- Comprehensive coverage of sensing techniques, their applications, and comparative analysis.
- Comprehensive coverage of sensing challenges and techniques to overcome those while maintaining the performance metrics.
- A holistic view of the sensing solutions through device-circuit co-optimization.
- Description of emerging challenges and future research directions.

Organization and Overview of the Content

This book is organized in four chapters, each covering a major class of emerging NVMs. The content of each chapter is designed to provide necessary background to the readers new in the field as well as detailed discussions on state-of-the-art techniques for the experts in the topic.

Chapter 1 presents various classes of spintronic memory technologies, their operating mechanisms, and the sensing techniques. The complex trade-offs between error resiliencies, access latency, energy, and footprint are also drawn along with a comparative analysis.

Chapter 2 covers resistive RAM technology, operation, and sensing techniques keeping both storage and non-storage applications in mind such as processing in memory and neuromorphic computing.

Chapter 3 describes ferroelectric capacitor- and ferroelectric transistor-based memory technologies and their sensing principles. A detailed comparative analysis is also presented.

Finally, Chap. 4 presents phase-change memory technology, their operating mechanism, and sensing principles.

Since several of the above memory technologies fall under resistive class, some of the sensing techniques can be suitable for more than one memory technology. Such similarities are pointed out wherever applicable to bring more clarity to the subject.

We believe the target audience consisting of students, researchers, and industry professionals will like the content and will be greatly benefited from it. We also believe that the content will remain highly relevant to other classes of emerging memories and alternate flavors of the memory technologies that are covered in this book. The content of this book is also relevant for new computing paradigms such as intelligent memories with compute capability and neuromorphic computing that are drawing significant push from industry and academia.

State College, USA   Swaroop Ghosh

Acknowledgements

We are very grateful to all the outstanding contributors for providing the high-quality chapters for the book. We highly appreciate the contributors for their wholehearted cooperation throughout this project. We also acknowledge the help and support from the students from the LOGICS Lab. We also sincerely thank Springer, USA, for support and all their publication materials.

We gratefully acknowledge National Science Foundation (NSF) under Grant CNS-1722557, Grant CNS-1814710, Grant CCF-1718474, and Grant DGE-1723687, and DARPA Young Faculty Award under Grant D15AP00089 for support. The views presented in this book are author's own and do not reflect the views of NSF and DARPA.

Contents

Editor and Contributors

About the Editor

Swaroop Ghosh (SM'13) received the B.E. (Hons.) degree from IIT Roorkee, India, in 2000; the M.S. degree from the University of Cincinnati in 2004; and the Ph.D. degree from Purdue University in 2008. From 2016, he is an Assistant Professor at Pennsylvania State University. Earlier, he was with the faculty of University of South Florida (USF) from 2012 to 2016. Prior to that, he was a Senior Research and Development Engineer in Advanced Design, Intel Corp, from 2008 to 2012. His research interests include next-generation circuits, cybersecurity, and digital testing. He holds 9 patents and has published over 90 refereed papers in this area.

He served as Associate Editor of the IEEE Transactions on Circuits and Systems I and as Senior Editorial Board member of IEEE Journal of Emerging Topics on Circuits and Systems (JETCAS). He has also served in the technical program committees of ACM/IEEE conferences such as DAC, ICCAD, CICC, DATE, ISLPED, ICCD, GLSVLSI, Nanoarch, and ISQED. He served as Program Chair of ISQED 2019 and DAC Ph.D. Forum 2016 and Track (co)-Chair of CICC 2017–2018, ISLPED 2017–2018, and ISQED 2016–2017.

He is a recipient of Intel's Technology and Manufacturing Group Excellence Award in 2010, Intel's Divisional Recognition Award in 2011, Intel's Departmental Recognition Award in 2011 and 2012, USF Outstanding Research Achievement Award in 2015, College of Engineering Outstanding Research Achievement Award in 2015, DARPA Young Faculty Award (YFA) in 2015, ACM SIGDA Outstanding New Faculty Award in 2016, YFA Director's Fellowship in 2017, Monkowski Early Career Award from the College of Engineering, Penn State in 2018, and Lutron Spira Teaching Excellence Award in 2018.

Contributors

Ahmedullah Aziz School of Electrical and Computer Engineering, Purdue University, West Lafayette, IN, USA

Suman Datta Department of Electrical Engineering, University of Notre Dame, South Bend, IN, USA

Sumitha George School of Electrical Engineering and Computer Science, Pennsylvania State University, State College, PA, USA

Swaroop Ghosh Pennsylvania State University, State College, PA, USA

Sumeet Kumar Gupta School of Electrical and Computer Engineering, Purdue University, West Lafayette, IN, USA; School of Electrical Engineering and Computer Science, Pennsylvania State University, State College, PA, USA

Rashmi Jha University of Cincinnati, Cincinnati, OH, USA

Alexander Jones University of Cincinnati, Cincinnati, OH, USA

Mohammad Nasim Imtiaz Khan Pennsylvania State University, State College, PA, USA

Hai Li Department of Electrical and Computer Engineering, Duke University, Durham, NC, USA

Xueqing Li School of Electrical Engineering and Computer Science, Pennsylvania State University, State College, PA, USA

Farshad Moradi Aarhus University, Aarhus, Denmark

Vijaykrishnan Narayanan School of Electrical Engineering and Computer Science, Pennsylvania State University, State College, PA, USA

Sandeep Krishna Thirumala School of Electrical and Computer Engineering, Purdue University, West Lafayette, IN, USA

Danni Wang School of Electrical Engineering and Computer Science, Pennsylvania State University, State College, PA, USA

Bonan Yan Department of Electrical and Computer Engineering, Duke University, Durham, NC, USA

Qing Yang Department of Electrical and Computer Engineering, Duke University, Durham, NC, USA

Behzad Zeinali Aarhus University, Aarhus, Denmark

Chapter 1
Sensing of Spintronic Memories

Behzad Zeinali and Farshad Moradi

1.1 Introduction

Miniaturization of the CMOS technology in the past few decades has enabled the benefits of higher performance, lower dynamic power, and larger integration density for memory arrays. However, further downscaling of the technology (e.g., 22 nm) has faced a big challenge due to the increasing process variations, short-channel effects (SCE) and leakage paths [1]. These cause the leakage current to become the major contributor to power consumption in memory arrays realized in scaled technology nodes [2]. The increasing leakage current along with the rising current density in silicon challenges manufacturers due to the excess heat generated by power dissipation, which will decrease the reliability and life span of memories. It means a great deal of the power spent is wasted solely on heat generation [3]. To deal with this challenge, different solutions at device, circuit, and architecture levels have been developed [4–13]. Although these contributions alleviate leakage current in comparison with the standard devices and memory architectures, the standby current is still challenging in high-performance memories.

This challenge can be addressed by the use of nonvolatile memories. Among different implementations of nonvolatile memories, magnetic random access memories (MRAM) have attracted significant attention due to the fact that they enable fast, high-density, and low-power memory architectures [14]. Conventionally, write operation of MRAMs has been challenging due to their higher dynamic energy. However, in scaled technology nodes, the reliability of read operation becomes a more important issue. Therefore, in this chapter, we discuss various MRAM technologies, sensing approaches for MRAMs, and advanced techniques to reliably read out data in scaled technology nodes. In Sect. 1.2, various families of MRAM memory such as STT-MRAM, SOT-MRAM, and multilevel MRAMs will be briefly described.

B. Zeinali · F. Moradi (✉)
Aarhus University, Aarhus, Denmark
e-mail: moradi@eng.au.dk

© Springer International Publishing AG, part of Springer Nature 2019
S. Ghosh (ed.), *Sensing of Non-Volatile Memory Demystified*,
https://doi.org/10.1007/978-3-319-97347-0_1

1

The advantages and disadvantages of each memory will be studied. In Sect. 1.3, different techniques to read out the data of MRAM cells will be investigated. In the first part of this section, advanced sense amplifier techniques will be studied, where the improvement in read operation is achieved by either removing the offset of the sense amplifier or increasing the distance between data voltage and reference voltage. The second part of Sect. 1.3 will devote to self-reference sensing methods. These methods will be categorized into destructive and nondestructive schemes, and in each category, several methods will be explained. This chapter will be concluded in Sect. 1.4.

1.2 Magnetic Random Access Memory (MRAM)

To achieve a memory with lower leakage and better reliability, there is a need for highly reliable nonvolatile memories. MRAM is a promising candidate due to its unique features including nonvolatility, long endurance, CMOS-compatibility, and short access time [15]. The MRAM uses magnetic tunneling junction (MTJ) technology as shown in Fig. 1.1, which has two permanent states with different resistance values, which can be utilized to store binaries, i.e., "0" or "1". The MTJ is composed of two magnetic layers, i.e., pinned layer (PL) and free layer (FL), and one oxide barrier layer. The PL is magnetically pinned, whereas the FL is not. Hence, it is easier to switch the FL rather than the PL. MTJ resistance is determined by the relative magnetization direction of two ferromagnetic layers. When magnetization directions of two magnetic layers are parallel (P-state) or antiparallel (AP-state), MTJ resistance is in low or high states, respectively. Thus, the MTJ can be used as a binary memory cell, low resistance (R_L) as logic "0" and high resistance (R_H) as logic "1". In addition, MTJs enable great scalability due to the fact that the critical switching current (I_C) is scaled in proportion to the size of the MTJ [16].

Based on easy axis direction, MTJs can be categorized into either in-plane-anisotropy MTJ (i-MTJ) or perpendicular-anisotropy MTJ (p-MTJ) as shown in

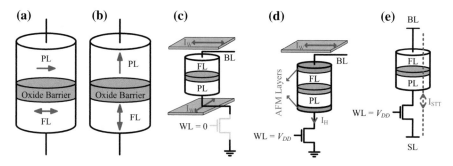

Fig. 1.1 **a** i-MTJ, **b** p-MTJ, **c** toggle switching, **d** TAS switching, and **e** 1T-1MTJ STT-MRAM

Fig. 1.1a, b respectively. In the case of i-MTJ, the thermal stability (Δ) is limited by the shape of MTJ due to the requirement of an elliptical shape to stabilize the magnetization along the in-plane axis in order to minimize the magnetostatic energy. However, keeping the elliptical shape and preventing magnetization curling due to thermal fluctuations are almost impossible in scaled technology nodes [17–26]. The p-MTJ can alleviate the size and shape limitations leading to higher density memory arrays with higher thermal stability [27]. In addition, the anisotropy and demagnetization fields are collinear in p-MTJs in contrary to the i-MTJs. Therefore, to enable switching of the magnetization direction of the FL in i-MTJs, we need to overcome a higher effective magnetic field leading to higher I_C and lower switching efficiency [27]. Hence, p-MTJs are more suitable in scaled technology nodes.

Several methods have been developed to switch between the two MTJ resistance states [28–31]. For the first generation of the MRAMs, switching between the two states is ensured by a magnetic field generated by two conductive lines as shown in Fig. 1.1c. In this configuration, the access transistor turns on only during read operation. A bidirectional current (I_W) flows through conductive lines, while the MTJ intersects the generated magnetic field. In case the magnetic field is strong enough, the magnetization direction of the FL will be switched. This switching mechanism requires two large currents that makes it questionable in terms of power efficiency. Furthermore, scaling the cell size will raise the data retention issue. To this end, thermally assisted (TAS) approach shown in Fig. 1.1d has been developed, which improves the field-induced magnetic switching approach. During write, a heating current (I_H) passes through the access transistor to heat up the MTJ above the blocking temperature of the FL. Then, it facilitates the switching process of the FL by an applied magnetic field. Therefore, according to Fig. 1.1d, TAS provides the benefit of using only one conductive line and one high switching current, while the low heating current allows a significant power saving. Furthermore, scalability and data retention are feasible using TAS method. It is worth mentioning that the MTJ is required to be modified to improve its thermal stability in TAS. This can be done by adding two antiferromagnetic (AFM) layers with different blocking temperatures. However, TAS still suffers from high write latency and it requires a high current to generate the magnetic field.

1.2.1 Spin-Transfer Torque (STT) MRAM

The widely used mechanism in scaled technologies only recruits a spin-polarized bidirectional current to apply an STT for MTJ switching as shown in Fig. 1.1e. The spin-polarized current (I_{STT}) is generated by applying appropriate voltages to the bit-line (BL) and the source-line (SL), and turning on the access transistor by asserting the WL signal. When electrons flow from the PL to the FL, the PL polarizes incoming electrons. These electrons are polarized in the spin direction of the PL, and after flowing into the FL, their spin momentum is transferred to the FL. Therefore, an STT is asserted to the FL to align its magnetization with the PL (i.e., P-state).

On the other hand, when electrons flow from the FL to the PL, electrons entering the FL are not polarized and can have any spin direction. Electrons with the same spin direction as the PL are able to tunnel through the oxide barrier easily. However, electrons with the opposite spin polarization accumulate in the FL. These electrons transfer their spin to the FL that asserts a torque aligning the magnetization of the FL in AP-state with the PL. When electrons transfer their spin to the FL, their spin directions become aligned with the spin polarization of the PL. They may then tunnel across the oxide barrier easily. Therefore, the process of parallelizing the FL and PL is more efficient than the antiparallelizing process.

In STT-MRAM, the asymmetric switching is aggravated even further due to different write currents. When current flows from BL to SL, the gate–source voltage of the access transistor is equal to the supply voltage (V_{DD}) leading to an overdrive voltage of $V_{DD}-V_{th}$. On the other hand, when current flows from SL to BL, the overdrive voltage of the transistor is considerably degraded due to the fact that the MTJ functions as a source degeneration resistance. Besides, the body-biasing effect increases the threshold voltage of the transistor and hence degrades the write current further. The above phenomena result in an asymmetric write operation in STT-MRAM cells [32].

Along with asymmetric switching, the MRAM switching process is inherently stochastic and actual write time varies with a distribution having a very long tail [33]. Total switching time of the MTJ consists of incubation time and transit time. The incubation time is defined as the time needed for electrons to climb up the potential barrier in the MTJ, while the transit time denotes the time for electrons to descend the potential barrier to the other state [34–37]. The incubation time contributes to more than 90% of total write time, and stochasticity of the switching delay results from variations in the incubation time mainly emerged from thermal fluctuations [38]. This stochasticity of the switching time leads to variation in write time, even for a single cell [39, 40]. Hence, in order to guarantee a reliable switching process, write current must be sustained for duration much longer than those are required for an average write to complete [33]. Increasing the write current can improve the switching operation of STT-MRAM as shown in Fig. 1.2 at the cost of higher write power consumption [41]. According to this figure, increasing I_{STT} can decrease both mean and variation of the switching time. However, this can threaten the reliability of STT-MRAM cells in sub-22-nm technology nodes. It is mainly due to this fact that passing a high write current through the MTJ imposes severe stress on the oxide layer leading to the occasional MTJ barrier breakdown.

1.2.2 Spin–Orbit Torque (SOT) MRAM

The reliability issue due to high write current can be addressed by emerging SOT switching mechanism as shown in Fig. 1.3 [42–46]. This technology works based on Spin Hall effect (SHE) [47–49] and/or Rashba effect [50], where a charge current (I_{SOT}) flowing through a non-magnetic heavy metal (HM) exerts a torque to switch the

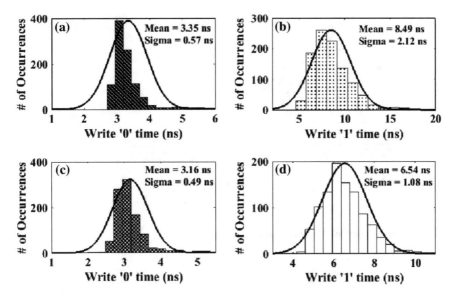

Fig. 1.2 Small access transistor in **a** write "0" and **b** write "1". Large access transistor in **c** write "0" and **d** write "1"

FL magnetization. Rashba effect originates from the breaking of structural symmetry [50, 51]. Structural asymmetry in magnetic heterostructures generates an interaction between electrical current and magnetic resulting a Rashba magnetic field (H_R). Therefore, in SOT-based MTJ devices where the FL is sandwiched between two different materials (HM and oxide layer) to break the vertical structural symmetry, passing I_{SOT} can generate an SOT induced by the Rashba magnetic field. The SOT switching can be also justified by SHE translated as the ability of generation of spin current from a non-spin-polarized charge current. In SHE, a charge current flowing through the heavy metal can generate spin accumulation on the lateral surfaces due to the strong spin–orbit interaction, which forms a pure spin current (I_s) with a polarization direction of $\vec{\sigma}_{SHE}$ [47–49, 52–55]. The SHE-induced spin current is injected into the adjacent FL generating an SOT to switch its magnetization direction. In comparison with the STT switching mechanism, SOT enables more energy efficient and faster write operation thanks to the higher effective spin injection efficiency [56]. It is shown that an effective spin injection efficiency >100% can be achieved by appropriate sizing of the HM [56].

Despite all the advantages of SOT-MRAMs, the main disadvantage of the conventional SOT-MRAM cell is area overhead, so that two transistors are required for each bit-cell leading to significant area overhead (30% overhead in [57]) in comparison with the STT-MRAM cell. Therefore, in high-density memory applications, SOT-MRAM cannot be an appropriate choice [57–59].

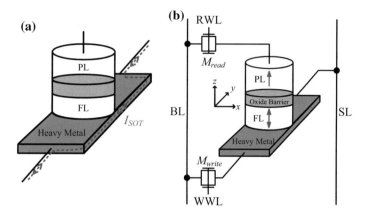

Fig. 1.3 **a** SOT-based MTJ, **b** SOT-MRAM cell

1.2.3 Multilevel MRAM

A well-known technique to alleviate the area overhead of memory cells is to implement a multilevel cell. The integration density of MRAM cells might be increased by storing multiple bits in each memory cell. This can be achieved by including more MTJs in each cell. In the following subsections, the promising multilevel MRAMs are briefly studied.

1.2.3.1 Multilevel SOT-MRAMs

In the case of SOT-MRAM, some multilevel designs have been recently developed [57, 59]. In [57], two series MTJ multilevel cell (S-MLC) and parallel MTJ multilevel cell (P-MLC) shown in Fig. 1.4a, b, respectively, are investigated and demonstrated. At S-MLC, two series MTJs make the storage element of the cell, while only the FL of MTJ_2 is in contact with the HM. According to this configuration (Fig. 1.4a), MTJ_2 can be switched by SOT mechanism, while MTJ_1 switching is accomplished by the STT current. During write operation, MTJ_1 is written first and then MTJ_2 is written that leads to write–disturb failure elimination in this cell. In this case, even if an unwanted switching occurs in MTJ_2 during programming of MTJ_1, the correct data can be written to MTJ_2 in the following cycle. However, it is obvious that during programming of MTJ_1, large write current flows through MTJ_2 as well, which degrades the reliability of MTJ_2. On the other hand, in P-MLC, the FLs of both MTJs are in contact with the HM and MTJs are connected in parallel to each other as shown in Fig. 1.4b. In this configuration, write operation of MTJs is accomplished by the SOT switching mechanism leading to no reliability issue. However, since both MTJs are in the same SOT current path, they should have different critical currents. It may be achieved by increasing cross-sectional area of the HM underneath the

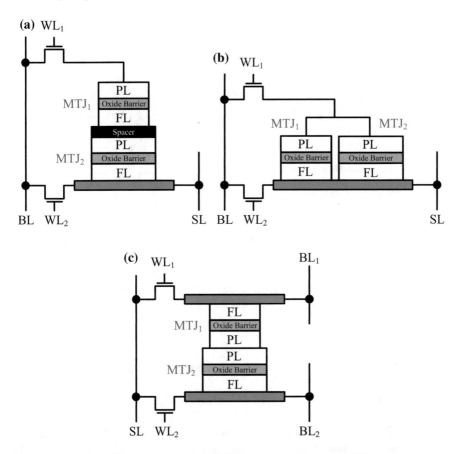

Fig. 1.4 **a** S-MLC, **b** P-MLC, **c** Multilevel SOT-MRAM cell with separated HMs

MTJ_2 leading to a higher I_C. In other words, by reducing the effective spin injection efficiency ($\frac{\eta A_{MTJ} I_{SOT}}{A_{HM}}$) of the SOT mechanism for MTJ_2, the requirement for different critical currents can be met. In the case of P-MLC, the MTJ with a larger I_C (MTJ_2) is programmed first and then the MTJ_1 is written by passing a lower SOT current than the critical current of the MTJ_2.

In [59], another multilevel SOT-MRAM cell has been developed, where the storage element contains a series stack of two MTJs with a shared PL as shown in Fig. 1.4c. In this cell, FLs of MTJ_1 and MTJ_2 are in contact with separate non-magnetic HMs enabling separate SOT-based switching for each MTJ. During the write operation of MTJ_1, the write current flows through the top HM by asserting the WL_1 signal, while the WL_2 signal is off. It is worth to mention that when the MTJ_1 is written, the sneak current passing through the stack MTJ may be eliminated by floating BL_2. On the other hand, switching of the MTJ_2 is enabled by asserting the WL_2 signal, while the WL_1 signal is off and the BL_1 is floating.

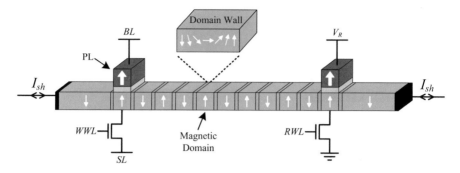

Fig. 1.5 DW racetrack memory

Read operation of the multilevel cells is enabled by providing four distinguishable resistance states. To this end, MTJs are realized with different cross-sectional areas; the MTJ_1 is designed with the minimum feature size of technology, while the MTJ_2 needs to be larger. Therefore, MTJ_2 requires a larger write current leading to higher write power dissipation. In addition, the MTJ_1 and MTJ_2 have different thermal stability due to different sizes.

1.2.3.2 Domain-Wall (DW) Racetrack Memory

This promising multilevel memory cell shown in Fig. 1.5 employs DW strip to store data, in which DWs are formed to separate magnetic domains of opposite magnetizations [60]. In addition, to write the data into and read it out from magnetic domains, write and read heads are utilized, respectively. Each head consists of a PL and an oxide barrier; hence, those can form an MTJ with their underneath magnetic domain. Therefore, each bit can be programmed or sensed by bringing the desired magnetic domain under the write or read head, respectively. It can be achieved by passing a charge current (I_{sh}) to shift magnetic domains in lockstep fashion. The shifting can be enabled when I_{sh} is larger than a threshold current (I_{th}), where I_{th} decreases with scaling the cross-sectional area of the strip. For $I_{sh} > I_{th}$, the velocity of moving DWs along the strip (v) can be represented as follows [61]:

$$v = \frac{\beta \mu P A_s}{\alpha q M_s}(I_{sh} - I_{th})$$

where β is the non-adiabatic coefficient, μ is the Bohr magneton, P is spin polarization, α is the damping factor, q is the elementary charge, and M_s is the saturation magnetization.

1.3 MRAM Sensing Techniques

Although, conventionally, write operation of MRAMs has been challenging due to higher dynamic energy, technology scaling can alleviate this challenge mainly due to scaling I_C proportional to the MTJ size. On the other hand, it also scales thermal stability (Δ) of the MTJ, which leads to a higher probability of undesirable switching during read operation. Therefore, it is challenging to reduce I_C while keeping a large Δ leading to a high energy overhead. This issue degrades the performance and reliability of read operation at scaled technologies. In scaled technology nodes, read disturbance which occurs when stored data into the MTJ is flipped increases mainly due to scaling of I_C. The read current should be sufficiently lower than the critical current, I_C, to avoid any disturbance during read operation. Although small read current alleviates the read disturbance, it will reduce the read margin, which considerably increases decision failure probability. The read margin is defined as the difference between BL voltage and reference voltage, and a good design with large read margin is required due to small tunneling magnetoresistance ratio (TMR) of available MTJ technologies.

1.3.1 Conventional Sense Amplifier

In the conventional read structure shown in Fig. 1.6, stored data of the MRAM cell is read out by comparing the MTJ resistance with a reference value [62]. The sensing current is generated in the reference branch and conveyed to the data branch through a current mirror circuit. The sensing current can be controlled by the clamp voltage (V_{clamp}) applied to the gate of clamp NMOS transistors. The clamp and current mirror circuits can be the main sources of process variation of the sensing circuit. The clamp transistors have source degeneration resistances, which reduce the variation effects. Furthermore, the degeneration PMOS transistors are included to keep the current through the current mirror circuit as fixed as possible. To improve the read margin, the reference voltage (V_{ref}) is selected between $V_{\text{data},H}$ and $V_{\text{data},L}$. These are data voltages (V_{data}) when the MTJ is at high and low resistance states, respectively. Since the sensing scheme is shared between multiple cells, the reference voltage must be chosen between the maximal $V_{\text{data},L}$ and the minimal $V_{\text{data},H}$. When the variation of the MTJ resistance is large, the two resistance states of MTJs can go higher or lower than the reference value. Therefore, finding a valid reference voltage for all involved MRAM cells is almost impossible due to large bit-to-bit variation of MTJ resistance leading to a high decision failure probability in scaled technology nodes. According to [62], the conventional structure can guarantee a sufficient sensing yield in 65-nm CMOS technology. However, by shrinking the technology node down to 45 nm, the sensing amplifier cannot provide a reliable read operation. To deal with the mentioned obstacle, advanced sensing circuits and new sensing schemes have been developed in scaled technology nodes.

Fig. 1.6 Conventional
sensing circuit of MRAM
cells

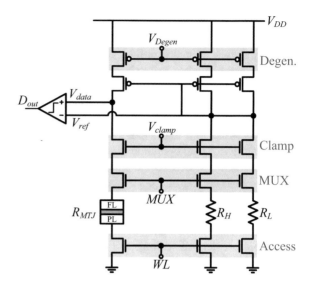

1.3.2 Error-Tolerant Sense Amplifiers

To improve the read reliability of MRAMs, the conventional sensing circuit in Fig. 1.6 can be replaced advanced sensing circuits [62–74]. In the conventional sensing circuits, there are two branches: data branch and reference branch generating V_{data} and V_{ref}, respectively. By scaling the technology, the distributions of both V_{data} and V_{ref} become wider leading to significant degradation of the sensing margin. Therefore, the reduction in the reference resistance distribution can improve the read margin. To enable this improvement, in [63], a multiple-cell reference scheme has been developed as shown in Fig. 1.7a, where $N \times N$ reference cells are utilized to reduce the variation of the reference resistance. According to Fig. 1.7a, this array structure is composed of MTJ resistances and on-state resistance of access transistors (R_{on}) generating an overall reference resistance (R_{ref}). The standard deviation of the reference resistance ($\sigma(R_{\text{ref}})$) can be expressed as follows [63]:

$$\sigma(R_{\text{ref}}) = \frac{\sqrt{\sigma(R_L)^2 + \sigma(R_H)^2 + 2\sigma(R_{\text{on}})^2}}{\sqrt{2}N}$$

where $\sigma(R_L)$, $\sigma(R_H)$, and $\sigma(R_{\text{on}})$ are the standard deviation of R_L, $R_{H,}$ and R_{on}, respectively. Therefore, by increasing the size of the array, read yield degradation attributed to the distribution of V_{ref} can be significantly improved.

Along with the increase in the standard deviations of V_{data} and V_{ref} in deep submicron technologies, due to supply voltage scaling, the difference between the mean values of V_{data} and V_{ref} is small. This is further aggravated by decreasing the critical current which imposes of using a low sensing current to prevent read disturbance. For instance, in 45-nm CMOS technology, the difference shrinks to 300 mV by consider-

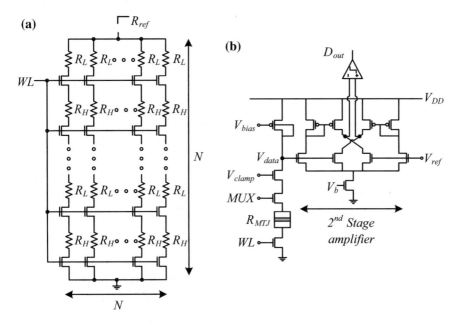

Fig. 1.7 **a** Multiple-cell reference scheme and **b** body-voltage-sensing-based short-pulse reading circuit

ing a very precise V_{ref} and using a sensing current of 26.6 μA [62]. In addition, using small current challenges the viability of a high-speed sensing scheme. To improve both the read margin and read speed, one solution is to use a short-pulse sensing scheme [64–66]. This idea has emerged from this fact that the switching probability of the MTJ depends on both the current amplitude and the duration of applying the current. In the low current scenario, a small sensing current may be applied for a long read duration leading to a slow sensing. On the other hand, the short-pulse sensing scheme enables the read operation by using a current that is close in amplitude to the write current but with much shorter pulse duration. Although this scheme can improve both the read margin and the read speed, an efficient circuit implementation of the short-pulse sensing method is very challenging in scaled technology nodes. Figure 1.7b shows a circuit realization for short-pulse sensing method [66]. In this sensing circuit, a body-connected PMOS transistor is utilized to convert the sensing current (I_{Sense}) to V_{data}. The output impedance of the body-connected load is $1/g_{\text{mb}}$ where g_{mb} is the body transconductance. It can be easily shown that the output impedance of the body-connected configuration is larger than that of the diode-connected load, while it is smaller than the output impedance of the current source load [66]. Therefore, it can enable a high-speed sensing operation with a proper sensing margin in comparison with other active load configurations. Besides, in Fig. 1.7b, the second stage amplifier generates extra signal swing at the output,

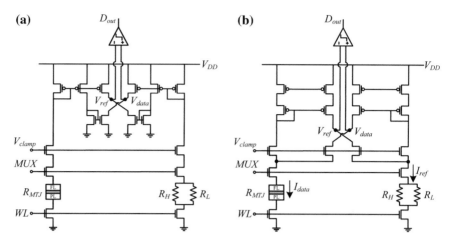

Fig. 1.8 Schematics of **a** HSCC, and **b** SPSC

enabling a more reliable and faster sensing operation. However, as shown in [62], V_{data} generated by the body-connected load is not still high enough to guarantee a sufficient sensing yield.

Beyond the mentioned methods to reduce the distributions of V_{data} and V_{ref}, other techniques have been also developed to improve the read yield, which use either data-aware dynamic reference voltage or latch offset cancelation techniques. In the following subsections, some of these methods are briefly described.

1.3.2.1 Data-Aware Dynamic Reference Voltage

In this category of sensing techniques, V_{ref} is variable and changes based on the level of V_{data}. When data is "0", V_{ref} increases, while for "1" state, V_{ref} decreases. This dynamic reference can significantly improve the read margin. Figures 1.8 and 1.9 show schematics of some data-aware dynamic reference voltage sensing schemes.

The sensing circuit with a highly symmetric cross-coupled current mirror (HSCC) is illustrated in Fig. 1.8a [67]. When data is "0", the current in the data branch is higher than the current in the reference branch. Therefore, by using the current mirror configuration, V_{ref} decreases and the difference of V_{ref} and V_{data} can be ideally doubled. The reverse scenario can be occurred when data is "1". However, to generate both V_{ref} and V_{data}, the HSCC utilizes three current mirrors leading to a very sensitive sensing margin to process variations in scaled technology nodes.

A modified version of the HSCC has been developed in [62] where only one current mirror is utilized to minimize the distribution of V_{data}. In the split-path sensing circuit (SPSC) shown in Fig. 1.8 b, similar to the HSCC, V_{ref} increases or decreases based on I_{data} leading to a two times larger read margin than that of the conventional sensing circuit. In addition, due to the use of two clamp transistors, V_{clamp} should be lower

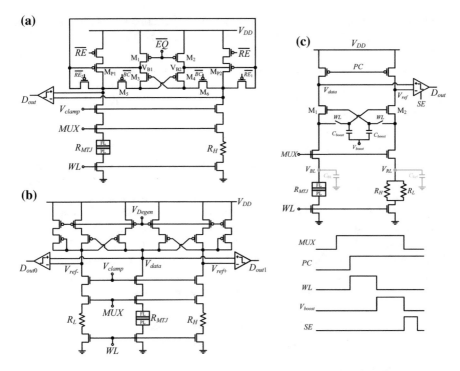

Fig. 1.9 Schematics of **a** DBSC, **b** DDRS, and **c** VFAB along with its operating phases

than that of the conventional sensing circuit to prevent read disturbance. Therefore, the currents passing through current mirrors are almost half of conventional sensing current leading to more sensitivity to process variations. Moreover, the sensing speed is asymmetric due to this fact that I_{data} is different during sensing "0" and "1" states.

To address the mentioned issues, the data-dependent body-bias sensing circuit (DBSC) has been developed in [68]. As shown in [68], this technique can provide a reliable sensing operation in deep sub-micrometer nodes, e.g., 45-nm CMOS technology. The schematic of the DBSC is shown in Fig. 1.9a, where six PMOS transistors (M_1–M_6) form a latch to control the body voltages of the PMOS load transistors (M_{P1} and M_{P2}). Before the sensing operation, the body terminals of the PMOS loads are precharged to the supply voltage, while during the sensing operation, based on the current in the data branch, the latch operates and the voltage of body terminals changes. For instance, when data is "0", since V_{data} is lower than V_{ref}, higher current passes through M_4 in comparison with that of M_3. Therefore, V_{B2} is discharged faster than V_{B1}; hence, the threshold voltage of M_{P2} will be lower than that of the M_{P1}. This means V_{ref} is charged through the corresponding diode-connected PMOS load (M_{P2}); hence, sensing margin improves. On the other hand, during sensing "1", the reverse scenario will happen and V_{data} increases. Although in this case V_{data} increases, read disturbance can be prevented by using proper configuration for the

MTJ. Since, during read operation, the unwanted switching may only happen in one direction (e.g., P to AP), the proper configuration of the MTJ (PL in contact with BL) can prevent the read disturbance. However, the main disadvantage of the technique relates to using the body-biasing approach to improve the sense margin. Some of the beyond CMOS technologies use thin body structures to improve short-channel effects [10]. In most of these technologies, adjusting the driveability of the transistor by body-biasing technique is not feasible.

Another dynamic reference sensing technique has been recently developed in [69] called dynamic dual-reference sensing (DDRS) scheme. As shown in Fig. 1.9b, the DDRS uses two complementary reference signals, V_{ref+} and V_{ref-}, generated by two reference cells with different reference resistance of R_H and R_L, respectively. In addition, different active loads are employed in each branch leading to dynamic reference voltages depending on V_{data}. For instance, during sensing "0", the negative feedback decreases the active loads of the reference branches and increases V_{ref+} and V_{ref-}. On the other hand, this feedback increases the active load of the data branch decreasing V_{data}. When the stored data is "1", the reference voltages decrease because of the negative feedback, while V_{data} increases. To enable the described process, the following conditions should be satisfied by the proper design of the active loads:

$$\text{Sensing ''0'' : } V_{data} < V_{ref-} < V_{ref+}$$
$$\text{Sensing ''1'' : } V_{ref} - < V_{ref}+, V_{data}$$

where satisfaction of these criteria can be challenging in scaled technology nodes. According to results in [69], the DDRS can provide $7\times$ larger sense margin in comparison with the conventional sensing circuit in 40-nm CMOS technology. Similar to the DBSC, the unwanted switching due to increasing V_{data} during sensing "1" can be addressed by proper configuring of the MTJ.

As the last technique in this section, voltage feedback and boosting (VFAB) scheme shown in Fig. 1.9c is reviewed [70]. The sensing operation starts by enabling precharge signal (PC) to charge two capacitors of C_{BL} and C_{Ref} to V_{DD}, where these are the parasitic capacitances of bit-line and reference-line, respectively. Then, WL is activated and C_{BL} and C_{Ref} start discharging through R_{MTJ} and $R_{Ref} = (R_H + R_L)/2$, respectively, leading to different discharging time constants. Due to different time constants, bit-line voltage (V_{BL}) and reference-line voltage (V_{RL}), which are fed to the gate of M_2 and M_1, respectively, alter with different rates. Therefore, the positive feedback between M_1 and M_2 can be enabled and one of these transistors goes to OFF state, while the other transistor stays in ON state. Afterward, the on-transistor can discharge the corresponding input voltage of the comparator (V_{data} or V_{ref}) and the data can be read out. For instance, when $R_{MTJ} = R_H$, V_{RL} is discharged faster than V_{BL} and M_1 turns off. Therefore, M_2 starts to discharge V_{ref} while $V_{data} = V_{DD}$. To speed up the discharge of V_{ref} (V_{data}) node, a boosting technique is employed at the gate of M_1 and M_2, which can improve the drivability of M_2 (M_1) by increasing the gate–source voltage of the transistor. To enable the boosting operation, WL is disabled and the boosting signal (V_{boost}) is applied leading to an increase in the

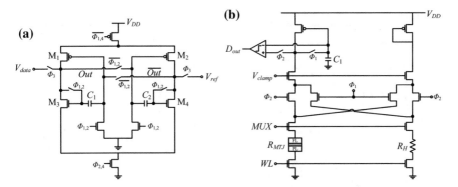

Fig. 1.10 Schematics of **a** OC-VLSA, and **b** TDSC

gate–source voltage of both M_1 and M_2. However, by careful selection of C_{boost} and V_{boost}, the V_{GS} of one transistor goes higher than its threshold voltage, while for the other transistor, it stays lower than its threshold voltage. According to the above description, although VFAB can improve the read margin significantly, it is very sensitive to C_{boost} and V_{boost} values as well as process variations, which can lead to changing in the threshold voltage of the transistors.

1.3.2.2 Latch Offset Cancelation

As mentioned in the previous subsection, the dynamic reference approaches increase the sensing margin to overcome process variations and improve the yield of the read operation. Another solution to improve the yield is to cancel the offset voltage of the sense amplifier. In this subsection, some techniques, working based on latch offset cancelation, will be reviewed.

Figure 1.10a shows the schematic of the offset cancelation voltage-latched sense amplifier (OC-VLSA) proposed in [71]. In this technique, during the preliminary phase, the mismatch between threshold voltages of the input transistors of the latch is sampled and stored in capacitors (C_1 and C_2). Then, during the latch phase, this mismatch is removed leading to a reliable read operation. This process is accomplished in four phases as follows:

1. Φ_1: C_1 and C_2 are precharged to V_{DD}.
2. Φ_2: C_1 and C_2 are discharged to the threshold voltages of M_3 and M_4, respectively.
3. Φ_3: V_{data} and V_{ref} are applied to the latch consisting M_1–M_4.
4. Φ_4: The decision is made regardless of the mismatch between latch transistors.

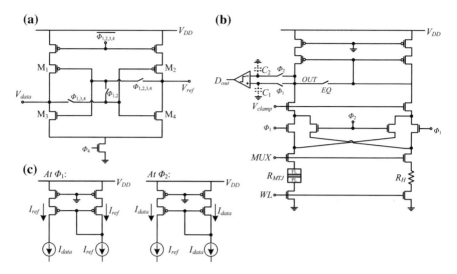

Fig. 1.11 Schematics of **a** LOC-SA and **b** OCDS-CS, and **c** the currents flowing through different branches of the OCDS-CS

However, obviously, this scheme suffers from slow sensing speed due to the requirement of three more phases and capacitances. In addition, the OC-VLSA only cancels the offset voltage due to M_3 and M_4, while mismatch between other transistors can also generate a significant offset voltage.

Another solution can enable offset cancelation in the latch, which has been developed in [72]. Figure 1.10b shows the schematic of the time differential sensing circuit (TDSC), which uses two phases for sensing. In the first phase (Φ_1), the reference cell is connected to the data branch and the output voltage is sampled in C_1. During the second phase (Φ_2), similar to the conventional sensing circuit, the output voltage is generated by the data cell. At the same time, the latch is activated to compare this voltage with the reference voltage generated in the Φ_1. Since the same PMOS load and clamp transistors are employed to generate V_{ref} and V_{data}, this scheme can eliminate the offset voltage emerged from this transistor in the conventional sensing circuit. However, similar to the OC-VLSA, using the capacitor and extra operating phase, degrade the read speed of the circuit.

The major source of the speed degradation in the above-mentioned approaches is the requirement of using additional capacitors. To address this issue, a capacitor-less latch offset cancelation sensing amplifier (LOC-SA) has been developed in [73] as shown in Fig. 1.11a. Similar to the OC-VLSA, this circuit works in four phases as follows where ΔV is defined as the difference between V_{data} and V_{ref}:

1. Φ_1: V_{data} and V_{ref} are equalized.
2. Φ_2: In this phase, V_{ref} is applied to the gate of M_1 and M_2. Therefore, similar to the conventional sensing circuit, V_{data} develops to $V_{ref} + \Delta V$.

3. Φ_3: In this phase, the same transistors in Φ_2 are utilized to compare and amplify ΔV with zero offset voltage. To enable it, V_{ref} is applied to the gate of M_1 and V_{data} ($V_{ref} + \Delta V$) is applied to the gate of M_2. Therefore, due to the positive feedback, ΔV is amplified and either V_{data} or V_{ref} is increased depending on the stored data in the MTJ. By increasing one of these voltages and due to the positive feedback, the other voltage decreases.

4. Φ_4: In this phase, the latch is activated and the decision is made. Since, during Φ_3, ΔV is developed, a fast decision can be generated by the latch.

This method can enable $2\times$ faster read operation in comparison with the TDSC. However, this technique still suffers from the offset voltage of other transistors, which can be challenging in scaled technology nodes.

To improve the read yield in deep sub-micrometer technologies, more innovative solution is a combination of offset cancelation and improving the sense margin. A possible circuit realization of this idea has been developed in [74] called offset-canceling dual-stage sensing circuit (OCDS-SC). Figure 1.11b shows the schematic of the OCDS-SC containing two operating phases as follows:

1. Φ_1: The output node (OUT) is connected to the data branch, and V_{data} is stored in C_1. At the beginning of the phase and for a short period, an equalization phase is conducted by activation of EQ signal. By doing this, the initial mismatch between data and reference branches is resolved, leading to sensing speed improvement.

2. Φ_2: The output node is connected to the reference branch, and V_{ref} is stored in C_2. Similar to the first phase, at the beginning of the second phase and for a short period, an equalization phase is conducted by activation of EQ signal.

In the OCDS-CS, similar to the TDSC, due to the use of the same pair of PMOS loads and clamp transistors to generate both V_{data} and V_{ref}, the offset voltage emerged from these transistors is removed. Along with the same transistors, V_{data} and V_{ref} are generated by the same current but by the opposite sign as shown in Fig. 1.11c. Therefore, the sense margin can be written as follows:

$$\Delta V = |V_{data} - V_{ref}| = |R_{load} \cdot (I_{ref} - I_{read}) - R_{load} \cdot (I_{read} - I_{ref})|$$
$$= 2R_{load} \cdot |I_{ref} - I_{read}|$$

where R_{load} is defined as the output resistance of the PMOS load. According to this analysis, the OCDS-CS can enable an offset cancelation scheme with a doubled sense margin. However, since V_{data} and V_{ref} are stored in parasitic capacitances of the latch input transistors, the reliability of the method at the present of process variations may be challenging.

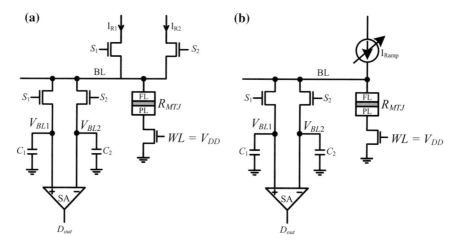

Fig. 1.12 Schematics of **a** conventional self-reference sensing scheme and **b** slope detection sensing scheme. In this scheme, S1 is activated before possible switching, while the switching may happen during S2

1.3.3 Self-Reference Sensing Scheme

Along with using advanced sensing circuits to improve the read margin and offset cancelation, self-reference read scheme has been proposed, which compares the original resistance state of the MTJ to a reference resistance generated by the same MTJ [75–78]. Such techniques can be categorized mainly into destructive and nondestructive sensing schemes. In destructive schemes, original data is destroyed during the read operation; hence, the data should be written back to the MTJ at the end of the read operation. On the other hand, nondestructive approaches preserve the original data during the read operation. In the following subsections, some of the recent self-reference sensing methods are studied.

1.3.3.1 Destructive Sensing Scheme

Under destructive self-reference sensing, the first scheme was proposed in [75] shown in Fig. 1.12a to overcome the bit-to-bit variation of the MTJ resistance. This technique consists of four steps: at the first step, a read current (I_{R1}) generates a BL voltage ($V_{BL1} = I_{R1} \times R_{MTJ}$) stored in C_1. Then, by conducting a write operation, data "0" is written into the MTJ, where this data is employed as the reference to determine the original data. At the third step, a larger read current (I_{R2}) flows through the MTJ to generate a BL voltage ($V_{BL2} = I_{R2} \times R_L$), which is stored in C_2. By a well-selected I_{R2}, the following relation can be expressed:

$$I_{R1} \times R_L < I_{R2} \times R_L < I_{R1} \times R_H$$

Therefore, by comparing the stored voltages in C_1 and C_2, the original data of the MTJ can be read out; when $V_{BL1} < V_{BL2}$, the original data is "0"; otherwise, it is "1". At the final step, the original data must be written back to the MTJ.

According to the above description, this sensing method requires two write operations leading to significant reliability degradation and a high power consumption. In addition, as shown in [76], at the presence of process variations, sensing margin is significantly limited due to the use of two sensing currents (I_{R1} and I_{R2}) leading to a high decision failure probability.

To eliminate the need of using I_{R1} and I_{R2}, a sensing scheme has been developed in [76] as shown in Fig. 1.12b. This method emerges from the fact that if the MTJ is in the AP-state, it will only switch to the P-state with a current flowing from BL to SL. In other words, when current flows from BL to SL, similar to what happens in read operation, the MTJ may be only switched from the AP-state to the P-state. When this occurs, the MTJ resistance state switches from high to low leading to a sudden voltage drop on the BL. Therefore, in [76], a ramp current is employed during the read operation, which results in a sudden drop in BL voltage with increasing sensing current if the MTJ resistance is switched from high to low state. In case this happens, the stored data is "1", else the data is "0". Although this method can improve the decision failure probability at the presence of process variations, it is still a destructive method and a write-back operation is required to restore the data "1" into the MTJ. Therefore, it will result in high power consumption and long access latency. Besides, similar to other destructive sensing schemes, since the process of writing back the data may require to be repeated several times to restore the original data into the MTJ, it can degrade the reliability of the whole memory.

1.3.3.2 Nondestructive Sensing Scheme

Nondestructive schemes are recently considered due to the fact that they do not need power-consuming write-back operation. However, they usually suffer from degraded sensing margin in more scaled technology nodes. In the following subsections, first, the nondestructive self-reference sensing scheme (NSRS) is explained [77]. Then, the nondestructive bit-line discharging (NBLD) scheme, which has been recently proposed by the authors, is described, and its efficiency is proved by extensive simulations [78].

Nondestructive Self-Reference Sensing Scheme (NSRS)

The NSRS is developed based on the different current roll-off slopes of the MTJ resistance in high and low states [77]. One of the interesting features of the MTJ resistance is that R_H and R_L for MgO-based MTJs provide quite different dependencies on the current; i.e., by increasing the amplitude of the current, R_H decays much faster than R_L as shown in Fig. 1.13a [78].

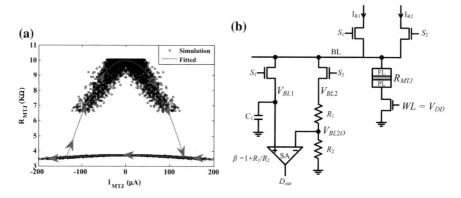

Fig. 1.13 **a** Static R–I curve of an MgO-based MTJ, **b** NSRS schematic

The NSRS includes two sensing steps as shown in Fig. 1.13b, but it eliminates the requirement of costly write steps in comparison with the destructive self-reference sensing schemes. The NSRS starts by asserting a read current (I_{R1}) to generate the BL voltage (V_{BL1}), which is stored on a capacitor (C_1). Then, the second read current (I_{R2}) is applied to generate the BL voltage (V_{BL2}), where $I_{R2} > I_{R1}$. The V_{BL2} is connected to a voltage divider to generate $V_{BL2O} = V_{BL2}/\beta$, where $\beta = I_{R2}/I_{R1}$. The sensing is accomplished by applying V_{BL1} and V_{BL2O} to the voltage sense amplifier. When V_{BL1} is significantly larger than V_{BL2O}, the stored data is "1"; otherwise, it is "0". This emerges from the fact that the MTJ resistance in AP-state under I_{R1} is much larger than that when I_{R2} is applied. On the other hand, when the data is "0", the MTJ resistance is almost constant under I_{R1} and I_{R2}; hence, $V_{BL1} \approx V_{BL2O}$. I_{R2} should be determined such that any disturbance in the MTJ-state is prevented, which can limit the read margin. However, in scaled technology nodes, sensing currents are also scaled, leading to a small resistance difference, when the current difference is subtle.

Nondestructive Bit-Line Discharging (NBLD)

To deal with the mentioned challenges in MTJ sensing, a nondestructive sensing scheme has been developed based on discharging of the precharged BL [78]. In this scheme, the BL capacitance is first precharged and then discharged during read operation. Since the sensing current is provided by the capacitor, the current amplitude decays over time. Therefore, to distinguish between high and low states of the MTJ resistance, different current roll-off slopes of the MTJ resistance in high and low states can be employed.

Figure 1.14 shows the circuit implementation of the NBLD sensing scheme and the signal waveforms for both read "0" and read "1". During sleep mode, the BL is precharged to V_{DD}. At the beginning of read, the SL is grounded and the BL

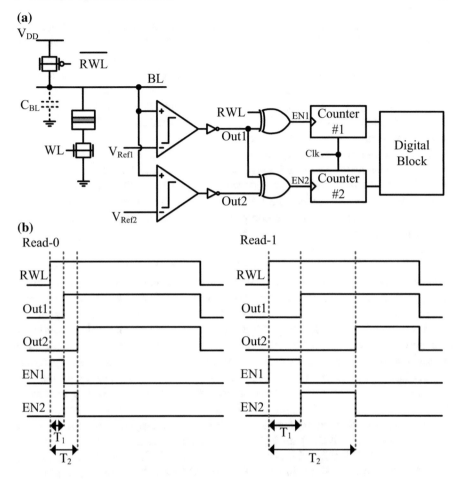

Fig. 1.14 **a** Schematic of nondestructive BL discharge circuitry and **b** signal waveforms for read "0" and read "1"

voltage (V_{BL}) decays due to the discharge of BL capacitance (C_{BL}). Therefore, V_{BL} is represented by the following equation where R_T consists of the MTJ resistance in series with the on-resistance of the access transistor:

$$V_{BL}(t) = V_{DD}\, e^{-\frac{t}{R_T C_{BL}}}$$

The sensing current, drawn from C_{BL}, starts from V_{DD}/R_T and decays to zero until the capacitance is fully discharged. Therefore, the MTJ resistance changes according to the right-hand side of the R–I curve in Fig. 1.13a. This is due to the fact that the sensing current changes exponentially. Therefore, when the stored data is "0", the MTJ resistance and consequently discharging time constant (τ_{BL}) of C_{BL} are almost constant. On the other hand, when the stored data is "1", the MTJ resistance increases by decreasing the current provided by C_{BL} during discharge. Therefore,

τ_{BL} increases during read "1". As a result, to read out the stored data, it should be determined whether the τ_{BL} is constant during BL discharge or not. When τ_{BL} is constant, the stored data is "0", and when it varies during BL discharge, the data is "1".

In the NBLD scheme, some features of the BL discharge are used to evaluate if the τ_{BL} is constant during the discharge or not. The BL starts to discharge from V_{DD} and decays to $V_{DD}\,e^{-1}$ and $V_{DD}\,e^{-2}$ at $T_1 = R_T C_{BL}$ and $T_2 = 2R_T C_{BL}$, respectively. Therefore, it can be easily concluded that when the stored data is "0", $T_1 \approx T_2 - T_1$, and when the data is "1", $T_1 < T_2 - T_1$.

In Fig. 1.14a, the BL voltage is applied to two sense amplifiers with different reference voltages of V_{Ref1} and V_{Ref2}, which are equal to $V_{DD}\,e^{-1}$ and $V_{DD}\,e^{-2}$, respectively. When the BL voltage crosses V_{Ref1} and V_{Ref2}, Out1 and then Out2 go high, respectively. Afterward, the EN1 and EN2 are generated and applied to the counters as enabling signals. By comparing the output of the counters, the stored data can be read out. The comparison can be accomplished by applying the outputs if the counter to the digital block.

The digital block can be implemented by XOR gates to compare the output of the counters. The main challenge in the proposed sensing scheme is to find an appropriate clock (*Clk*) frequency for the counters. To reduce the read latency, the sensing scheme needs a clock signal with low-jitter few-GHz frequency. To provide it, high-speed clock generators [79] or dual-edge triggered counters [80] can be used. According to the above description, the NBLD scheme has the benefit of sensing MTJ resistance states in one step and no penalty in latency. Besides, at the beginning of read operation, the sensing current may be higher than I_C; hence, the technique improves the read margin. It is mainly due to this fact that when read operation starts, V_{BL} is close to the supply voltage. Therefore, the sensing current is equal to that of write operation. It is worth to mention that for small values of C_{BL}, the charge on the capacitor is not enough to switch the MTJ-state, protecting it against read disturbance.

A 1T-1MTJ STT-MRAM bit-cell that uses NBLD during read operation is simulated in 14-nm FinFET technology. STT-MRAM bit-cell parameters used in this design are given in Table 1.1. With these parameters, the values for R_L and R_H are 3.46 and 10 KΩ, respectively, which result in a TMR of 189%. For the following simulations, the value of C_{BL} is chosen 200 fF. Although the reference voltages are selected $V_{DD}\,e^{-1}$ and $V_{DD}\,e^{-2}$, producing such precise voltages is impractical. Therefore, in simulations, reference voltages of the sense amplifiers are chosen as 0.28 and 0.1 V, respectively.

Figure 1.15 shows the control and the generated signals of the STT-MRAM during write and read operations. According to this figure, during read "1", the MTJ resistance changes from 6.5 to 10 KΩ, while, during read "0", the MTJ resistance is almost constant. In addition, this figure shows that the duration of write "1" and that of write "0" are 8.3 and 5.3 ns, respectively. In case of read, Tr11 and Tr01 variables are defined, which are widths of EN1 signal during read "1" and read "0" operations, respectively. Furthermore, Tr12 and Tr02 are defined as the widths of the EN2 signal during read "1" and read "0" operations, respectively. Therefore, read "1" latency (Tr1) can be calculated as Tr1 = Tr11 + Tr12 while for read "0", the latency (Tr0) is Tr0 = Tr01 + Tr02.

Table 1.1 STT-MRAM cell parameters

Parameter	Value
Free layer dimensions	25 nm × 25π nm × 1.4 nm
Oxide thickness (T_{MgO})	1.15 nm
Saturation magnetization (M_S)	700 emu/cm^3
Damping factor (α)	0.028
Gyromagnetic factor (γ)	17.6 MHz/Oe
Supply voltage (V_{DD})	0.75 V
Number of fins	10
Transistor length (L_g)	30 nm

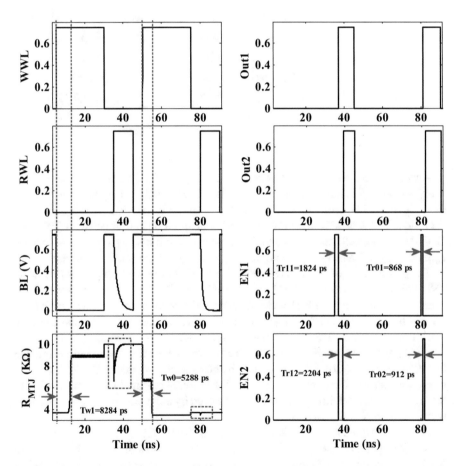

Fig. 1.15 Nondestructive BL discharging for read "1" and read "0"

Fig. 1.16 Distribution of
EN1 and EN2 time widths
during read "0" and read "1"

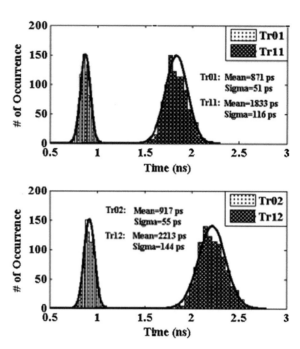

According to the simulation results, Tr1 and Tr0 are 4.03 ns and 1.78 ns, respectively. Besides, this figure shows that Tr01 and Tr02 are almost equal, while Tr12 is larger than Tr11 mainly due to the increase in the MTJ resistance during read "1" operation. It is worth to mention that there is a small deviation between Tr01 and Tr02 due to the reference voltages of the sense amplifiers, which are not exactly equal to $V_{DD} e^{-1}$ and $V_{DD} e^{-2}$ and the offset voltage of the sense amplifiers.

The performance of the NBLD sensing scheme in the presence of process variation and device mismatches is studied by the use of Monte Carlo simulations. Monte Carlo simulations for 14-nm FinFET technology include variations of the gate length (L_g), fin width (W_{fin}), fin height (H_{fin}), effective oxide thickness (*EOT*), and the gate work function. For all these parameters, the value of 3σ is equal to 10% of the nominal physical value. Furthermore, for MTJ, the Monte Carlo simulations include the variation of the oxide thickness and the cross-sectional area of the MTJ (A_{MTJ}), where the value of 3σ is equal to 10% of the nominal physical value. According to these quantities, the following results are achieved by doing Monte Carlo simulations with 1000 samples for each case.

Figure 1.16 illustrates the distribution of the Tr01, Tr02, Tr11, and Tr12 pulse widths during read "0" and read "1" operations. These distributions can be utilized to find the appropriate *Clk* frequency. The clock frequency must be chosen to guarantee that the counters show the same values for maximum pulse widths during read "0". On the other hand, it should show different values for minimum widths during read "1" operation. Here, 5σ is considered to find the worst-case scenario giving a bit

Table 1.2 Worst-case EN1 and EN2 time widths

Parameter	$T_{r01\text{-max}}$	$T_{r02\text{-max}}$	$\Delta T_{r0\text{-max}}$	$T_{r11\text{-min}}$	$T_{r12\text{-min}}$	$\Delta T_{r1\text{-min}}$
Value (ps)	1126	1192	66	1253	1493	240

Table 1.3 Comparison of different sensing approaches in scaled technology nodes

	I_{Sense}	Yield	Access time	E_{Sense}/bit	Area
Conventional SA	Low	Low	Fast	Low	Medium
Short-pulse sensing	High	Medium	Very fast	Low	High
HSCC	Low	Low	Fast	Medium	High
SPSC	Low	Medium	Medium	Medium	High
DBSC	Low	Medium	Fast	Low	Small
DDRS	Low	High	Fast	Medium	High
OC-VLSA	Low	Low	Slow	Medium	Medium
TDSC	Low	Medium	Medium	Low	Medium
LOC-SA	Low	High	Fast	Low	Medium
OCDS-SC	Low	High	Fast	Low	Medium
Self-reference sensing [75]	Low	Low	Very slow	Very high	High
Slope detection [76]	High	High	Very slow	Very high	Very high
NSRS	Low	High	Very slow	High	Very high
NBLD	High	High	Medium	Medium	Very high

error rate (BER) in the order of 10^{-6} during read operation. The corresponding values are shown in Table 1.2. According to this table, the minimum clock frequency is 2 GHz for dual-edge triggered counters. The clock frequency can be lowered when the C_{BL} and/or the demanding BER increase. Besides, according to Fig. 1.16, it can be concluded that the maximum read latency is 5.4 ns, which may occur when Tr11 and Tr12 are 5σ longer than their mean values.

At the end of this chapter, a comparison between the reviewed sensing techniques in scaled technology nodes is shown in Table 1.3. In this table, parameters such as I_{Sense}, sensing yield, access time, read energy, and area which are critical to evaluate the efficiency of sensing scheme are compared. According to this comparison, error-tolerant sense amplifiers can enable a fast, low power, and area-efficient sensing operation. However, these techniques generally suffer from increasing process variation. Therefore, their efficiency will be questioned in scaled technologies. On the other hand, self-reference techniques are more robust against variations of MTJ resistance leading to a higher reliability in cutting-edge technologies. However, this benefit is achieved at the cost of longer delay, higher power consumption, and more complexity.

Although the reviewed sensing approaches have been developed to improve the yield/sense margin of MRAMs/STT-MRAMs, the reader may realize that these can be easily employed on sensing of other spintronic memories, e.g., DW racetrack memories, SOT-MRAMs, magnetic skyrmions, spin memristor.

1.4 Conclusion

In this chapter, we reviewed different types of MRAMs, possible solutions to improve the density of bit-cells and different sensing techniques. In this respect, the MTJ as the storage element of MRAMs was briefly described and different methods for writing data into the MTJ were explained. Then, promising multilevel MRAM cells, i.e., multilevel SOT-MRAMs and DW racetrack MRAMs, were studied. After this background, different sensing schemes to read out the stored data in the MTJ were explained. Since the yield of read operation has been significantly degraded in recent cutting-edge semiconductor technologies due to large process variation and small read margin, new sensing schemes improving the yield of the memories have attracted significant attention. In this chapter, both error-tolerant sense amplifiers and self-reference sensing schemes are reviewed. In the case of error-tolerant sense amplifiers, different techniques to improve the read yield either by changing the reference voltage based on the data or removing the latch offset voltage were explored. Furthermore, for self-reference sensing, several techniques categorized as destructive or nondestructive were reviewed. According to our review, it becomes increasingly important and necessary to improve the reliability of MRAM cells in scaled technology nodes using such approaches.

References

1. Chau R, Doyle B, Datta S, Kavalieros J, Zhang K (2007) Integrated nanoelectronics for the future. Nature Mater 6(11):810–812
2. Roy K, Mukhopadhyay S, Mahmoodi-Meimand H (2003) Leakage current mechanisms and leakage reduction techniques in deep submicrometer CMOS circuits. Proc IEEE 91(2):305–327
3. Pal A (2014) Low-Power VLSI circuits and systems. Springer, New Delhi
4. Chen YH, Chan WM, Wu WC, Liao HJ, Pan KH, Liaw JJ, Chung TH, Li Q, Lin CY, Chiang MC, Wu SY, Chang J (2015) A 16 nm 128 Mb SRAM in high-k metal-gate finfet technology with write-assist circuitry for low-VMIN applications. IEEE J Solid-State Circuits 50(1):170–177
5. Raychowdhury A, Geuskens B, Kulkarni J, Tschanz J, Bowman K, Karnik T, Lu SL, De V, Khellah MM (2010) PVT-and-aging adaptive word-line boosting for 8T SRAM power reduction. In: IEEE ISSCC Digest of Technical Papers, pp 352–353
6. Verma N, Chandrakasan A (2008) A 256 kb 65 nm 8T sub-Vt SRAM employing sense-amplifier redundancy. IEEE J Solid-State Circuits 43(1):141–149

7. Hirabayashi O, Kawasumi A, Suzuki A, Takeyama Y, Kushida K, Sasaki T, Katayama A, Fukano G, Fujimura Y, Nakazato T, Shizuki Y, Kushiyama N, Yabe T (2009) A process-variation-tolerant dual-power-supply SRAM with 0.179 μm² cell in 40 nm CMOS using level-programmable wordline driver. In: IEEE Int. Solid-State Circuits Conference on Digest of Technical Papers, pp 458–459
8. Nii K, Yabuuchi M, Tsukamoto Y, Ohbayashi S, Oda Y, Usui K, Kawamura T, Tsuboi N, Iwasaki T, Hashimoto K, Makino H, Shinohara H (2008) A 45-nm single-port and dual-port SRAM family with robust read/write stabilizing circuitry under DVFS environment. In: VLSI Circuits Dig, pp 212–213
9. Ghanatian H, Hosseini SE, Zeinali B, Moradi F (2017) Quasi-Schottky-Barrier UTBB SOI MOSFET for low-power robust SRAMs. IEEE Trans Electron Devices 64(4):1575–1582
10. Rim K, Chan K, Shi L, Boyd D, Ott J, Klymko N, Cardone F, Tai L, Koester S, Cobb M, Canaperi D, To B, Duch E, Babich I, Carruthers R, Saunders P, Walker G, Zhang Y, Steen M, Ieong M (2003) Fabrication and mobility characteristics of ultra-thin strained Si directly on insulator (SSDOI) MOSFETs. In: Int. Electron Device Meeting (IEDM) Technical Digest, pp 47–52
11. Wilk GD, Wallace RM, Anthony JM (2001) High-K gate dielectrics: current status and materials properties considerations. J Appl Phys 89(10):5243–5275
12. Ota H, Hirano A, Watanabe Y, Yasuda N, Iwamoto K, Akiyama K, Okada K, Migita S, Nabatame S, Toriumi A (2007) Intrinsic origin of electron mobility reduction in high-K MOSFETs—from remote phonon to bottom interface dipole scattering. In: International Electron Device Meeting (IEDM) Technical Digest, pp 65–68
13. Frank DJ, Wong HSP (2000) Analysis of the design space available for high-k gate dielectrics in nanoscale MOSFETs. Superlattices Microstruct 28(5–6):485–491
14. Bagheriye L, Toofan S, Saeidi R, Zeinali B, Moradi F (2017) A reduced store/restore energy MRAM-Based SRAM cell for a non-volatile dynamically reconfigurable FPGA. In: IEEE transactions on circuits and systems II: Express briefs
15. Huai Y (2008) Spin-transfer torque MRAM (STT-MRAM): challenges and prospects. AAPPS Bull 18(6):33–40
16. Hosomi M, Yamagishi H, Yamamoto T, Bessho K, Higo Y, Yamane K, Yamada H, Shoji M, Hachino H, Fukumoto C, Nagao N, Kano H (2005) A novel nonvolatile memory with spin torque transfer magnetization switching: Spin-RAM. In: International Electron Device Meeting (IEDM) Technical Digest, pp 473–476
17. Mangin S, Ravelosona D, Katine JA, Carey MJ, Terris BD, Fullerton EE (2006) Current-induced magnetization reversal in nanopillars with perpendicular anisotropy. Nature Mater 5:210–215
18. Meng H, Wang JP (2006) Spin transfer in nanomagnetic devices with perpendicular anisotropy. Appl Phys Lett 88:172506
19. Kishi T, Yoda H, Kai T, Nagase T, Kitagawa E, Yoshikawa M, Nishiyama K, Daibou T, Nagamine M, Amano M, Takahashi S, Nakayama M, Shimomura N, Aikawa H, Ikegawa S, Yuasa S, Yakushiji K, Kubota H, Fukushima A, Oogane M, Miyazaki T, Ando K (2008) Lower-current and fast switching of a perpendicular TMR for high speed and high density spin-transfer-torque MRAM. In: International Electron Device Meeting (IEDM) Technical Digest, pp 309–312
20. Nishimura N, Hirai T, Koganei A, Ikeda T, Okano K, Sekiguchi Y, Osada Y (2002) Magnetic tunnel junction device with perpendicular magnetization films for high-density magnetic random access memory. J Appl Phys 91(8):5246–5249
21. Ohmori H, Hatori T, Nakagawa S (2008) Perpendicular magnetic tunnel junction with tunnelling magnetoresistance ratio of 64% using MgO(100) barrier layer prepared at room temperature. J Appl Phys 103:07A911
22. Yoshikawa M, Kitagawa E, Nagase T, Daibou T, Nagamine M, Nishiyama K, Kishi T, Yoda H (2008) Tunnel magnetoresistance over 100% in MgO-based magnetic tunnel junction films with perpendicular magnetic L10-FePt electrodes. IEEE Trans Magn 44(11):2573–2576
23. Kim G, Sakuraba Y, Oogane M, Ando Y, Miyazaki T (2008) Tunnelling magnetoresistance of magnetic tunnel junctions using perpendicular magnetization L10-CoPt electrodes. Appl Phys Lett 92:172502

24. Carvello B, Ducruet C, Rodmacq B, Auffret S, Gautier E, Gaudin G, Dieny B (2008) Sizable room-temperature magnetoresistance in cobalt based magnetic tunnel junctions with out-of-plane anisotropy. Appl. Phys. Lett 92:102508
25. Park JH, Park C, Jeong T, Moneck MT, Nufer NT, Zhu JG (2008) Co/Pt multilayer based magnetic tunnel junctions using perpendicular magnetic anisotropy. J. Appl. Phys 103:07A917
26. Mizunuma K, Ikeda S, Park JH, Yamamoto H, Gan H, Miura K, Hasegawa H, Hayakawa J, Matsukura F, Ohno H (2009) MgO barrier-perpendicular magnetic tunnel junctions with CoFe/Pd multilayers and ferromagnetic insertion layers. Appl Phys Lett 95:232516
27. Ikeda S, Miura H, Yamamoto H, Mizunuma K, Gan HD, Endo M, Kanai S, Hayakawa J, Matsukura F, Ohno H (2010) A perpendicular-anisotropy CoFeB–MgO magnetic tunnel junction. Nature Mater 9:721–724
28. Carpentieri M, Tomasello R, Ricci M, Burrascano P, Finocchio G. Micromagnetic study of electrical-field-assisted magnetization switching in MTJ devices. In: IEEE Transactions on Magnetics 50(11)
29. Fong X, Kim Y, Choday SH, Roy K (2014) Failure mitigation techniques for 1T-1MTJ spin-transfer torque MRAM bit-cells. IEEE Trans Very Large Scale Integ Syst 22(2):384–395
30. Jovanovic B, Brum RM, Torres L (2015) Comparative analysis of MTJ/CMOS hybrid cells based on TAS and in-plane STT magnetic tunnel junctions. In: IEEE Transactions on Magnetics 51(2)
31. He W, Zhu T, Zhang XQ, Yang HT, Cheng ZH (2013) Ultrafast demagnetization enhancement in CoFeB/MgO/CoFeB magnetic tunneling junction driven by spin tunneling current. Sci Rep 3:2883
32. Farkhani H, Peiravi A, Moradi F (2016) Low-energy write operation for 1T-1MTJ STT-RAM bitcells with negative bitline technique. IEEE Trans Very Large Scale Integ (VLSI) Syst 24(4):1593–1597
33. Zheng T, Park J, Orshansky M, Erez M. Variable-energy write STT-RAM architecture with bit-wise write-completion monitoring. In: Symposium on Low Power Electronics and Design, pp 229–234
34. Panagopoulos GD, Augustine C, Roy K (2013) Physics-based SPICE-compatible compact model for simulating hybrid MTJ/CMOS circuits. IEEE Trans Electron Devices 60(9):2808–2814
35. Guo W, Prenat G, Javerliac V, El Baraji M, De Mestier D, Baraduc C, Dieny B (2010) SPICE modeling of magnetic tunnel junctions written by spin-transfer torque. J Appl Phys 43:215001
36. Madec M, Kammerer JB, Pregaldiny F, Herbrard L, Lallement C (2008) Compact modeling of magnetic tunnel junction. In: Proceedings on 6th International IEEE Northeast Workshop on Circuits System TAISA Conference, pp 229–232
37. Iga F, Yoshida Y, Ikeda S, Hanyu T, Ohno H, Endoh T (2012) Time-resolved switching characteristic in magnetic tunnel junction with spin transfer torque write scheme. Jpn J Appl Phys 51
38. Devolder T, Hayakawa J, Ito K, Takahashi H, Ikeda S, Crozat P, Zerounian N, Kim JV, Chappert C, Ohno H (2008) Single-shot time-resolved measurements of nanosecond-scale spin-transfer induced switching: stochastic versus deterministic aspects. Phys Rev Lett 100(5)
39. Farkhani H, Peiravi A, Madsen JK, Moradi F (2015) STT-RAM write energy consumption reduction by differential write termination method. In: IEEE International Symposium on Circuits and Systems (ISCAS), pp 2936–2939
40. Farkhani H, Tohidi M, Peiravi A, Madsen JK, Moradi F (2017) STT-RAM energy reduction using self-referenced differential write termination technique. IEEE Trans Very Large Scale Integ (VLSI) Syst 25(2):476–487
41. Zeinali B, Karsinos D, Moradi F (2017) Progressive scaled STT-RAM for approximate computing in multimedia applications. IEEE Trans Circuits Syst II: Exp Briefs
42. Miron IM, Gaudin G, Auffret S, Rodmacq B, Schuhl A, Pizzini S, Vogel J, Gambardella P (2010) Current-driven spin torque induced by the Rashba effect in a ferromagnetic metal layer. Nature Mater 9:230–234

43. Miron IM, Garello K, Gaudin G, Zermatten PJ, Costache MV, Auffret S, Bandiera S, Rodmacq B, Schuhl A, Gambardella P (2011) Perpendicular switching of a single ferromagnetic layer induced by in-plane current injection. Nature Lett 476:189–194

44. Liu L, Lee OJ, Gudmundsen TJ, Ralph DC, Buhrman RA (2012) Current-induced switching of perpendicularly magnetized magnetic layers using spin torque from the spin Hall effect. Phys Rev Lett 109:096602

45. Liu L, Pai CF, Li Y, Tseng HW, Ralph DC, Buhrman RA (2012) Spin-torque switching with the giant spin Hall effect of tantalum. Science 336:555–558

46. Pai CF, Liu L, Tseng HW, Ralph DC, Buhrman RA. Spin transfer torque devices utilizing the giant spin Hall effect of tungsten. Appl Phys Lett 101(12)

47. Dyakonov MI, Perel VI (1971) Current-induced spin orientation of electrons in semiconductors. Phys Lett A 35(6):459–460

48. Hirsch JE (1999) Spin Hall effect. Phys Rev Lett 83(9)

49. Zhang S (2000) Spin Hall effect in the presence of spin diffusion. Phys Rev Lett 85(2):393–396

50. Bychkov YA, Rashba EI (1984) Properties of a 2D electron gas with lifted spectral degeneracy. J Exp Theor Phys Lett 39(2):78–81

51. Yu G, Upadhyaya P, Fan Y, Alzate JG, Jiang W, Wong KL, Takei S, Bender SA, Chang LT, Jiang Y, Lang M, Tang J, Wang Y, Tserkovnyak Y, Amiri PK, Wang KL (2014) Switching of perpendicular magnetization by spin–orbit torques in the absence of external magnetic fields. Nature Nanotechnol 9:548–554

52. Dyakonov MI, Perel VI (1971) Possibility of orienting electron spins with current. J Exp Theor Phys Lett 13(11):467–469

53. Jungwirth T, Wunderlich J, Olejník K (2012) Spin Hall effect devices. Nature Mater 11:382–389

54. Zeinali B, Madsen JK, Raghavan P, Moradi F (2017) Ultra-Fast SOT-MRAM Cell with STT current for deterministic switching. In: IEEE International Conference on Computer Design (ICCD), pp 463–468

55. Wang Z, Zhao W, Deng E, Klein JO, Chappert C (2015) Perpendicular-anisotropy magnetic tunnel junction switched by spin Hall-assisted spin-transfer torque. J Phys D Appl Phys 48(6):065001

56. Seo Y, Kwon KW, Roy K (2016) Area-efficient SOT-MRAM with a Schottky diode. IEEE Elect Dev Lett 37(8):982–985

57. Kim Y, Fong X, Kwon KW, Chen MC, Roy K (2015) Multilevel spin-orbit torque MRAMs. IEEE Trans Elect Dev 62(2):561–568

58. Seo Y, Fong X, Kwon KW, Roy K. Spin-Hall magnetic random-access memory with dual read/write ports for on-chip caches. IEEE Magn Lett 6

59. Zeinali B, Esmaeili M, Madsen JK, Moradi F (2017) Multilevel SOT-MRAM cell with a novel sensing scheme for high-density memory applications. In: 47th European Solid-State Device Research Conference (ESSDERC), pp 172–175

60. Ghosh S (2013) Design methodologies for high density domain wall memory. In: NANOARCH

61. Dong Q, Yang K, Fick L, Fick D, Blaauw D, Sylvester D. Low-power and compact analog-to-digital converter using spintronic racetrack memory devices. IEEE Trans Very Large Scale Integ (VLSI) Syst 25(3):907–918

62. Kim J, Na T, Kim JP, Kang SH, Jung SO (2014) A, split-path sensing circuit for spin torque transfer MRAM. IEEE Trans Circuits Syst-II Exp Briefs 61(3):193–197

63. Na T, Kim JP, Kang SH, Jung SO (2016) Multiple-cell reference scheme for reduced reference resistance distribution in deep submicrometer STT-RAM. IEEE Trans Very Large Scale Integ (VLSI) Syst 24(9):2993–2997

64. Ono K, Kawahara T, Takemura R, Miura K, Yamamoto H, Yamanouchi M, Hayakawa J, Ito K, Takahashi H, Ikeda S, Hasegawa H, Matsuoka H, Ohno H (2009) A disturbance-free read scheme and a compact stochastic-spin-dynamics-based mtj circuit model for Gb-scale SPRAM. In: Proceedings of IEEE International Electron Devices Meeting, pp 1–4

65. Ren F, Park H, Dorrance R, Toriyama Y, Yang CKK. A body-voltage-sensing-based short pulse reading circuit for spin-torque transfer RAMs (STT-RAMs). In: Proceedings of 13th International Symposium Quality Electronic Design (ISQED'12), pp 275–282

66. Ren F, Park H, Yang CKK, Markovic D. Reference calibration of body-voltage sensing circuit for high-speed STT-RAMs. IEEE Trans. Circuits Syst. I: Reg. Papers 60(11):2932–2939
67. Maffitt TM, DeBrosse JK, Gabric JA, Gow ET, Lamorey MC, Parenteau JS, Willmott DR, Wood MA, Gallagher WJ (2006) Design considerations for MRAM. IBM J Res Develop 50(1):25–49
68. Jo K, Yoon H (2017) Variation-tolerant sensing circuit for ultralow-voltage operation of spin-torque transfer magnetic RAM. IEEE Trans Circuits Syst II: Exp Briefs 64(5):570–574
69. Kang W, Pang T, Lv W, Zhao W. A dynamic dual-reference sensing scheme for deep submicrometer STT-MRAM. IEEE Trans Circuits Syst I: Reg Papers 64(1):122–132
70. Motaman S, Ghosh S, Kulkarni JP (2017) VFAB: A Novel 2-Stage STTRAM sensing using voltage feedback and boosting. IEEE Trans Circuits Syst I: Reg Papers
71. Javanifard J, Tanadi T, Giduturi H, Loe K, Melcher RL, Khabiri S, Hendrickson NT, Proescholdt AD, Ward DA, Taylor MA (2008) A 45 nm self-aligned-contact process 1 Gb NOR flash with 5 MB/s program speed. In: Proceeding of International Solid-State Circuits Conference, pp 424–426
72. Jefremow M, Kern T, Allers W, Peters C, Otterstedt J, Bahlous O, Hofmann K, Allinger R, Kassenetter S, Landsiedel DS (2013) Time-differential sense amplifier for sub-80 mV bitline voltage embedded STT-MRAM in 40 nm CMOS. In: Proceedings of International Solid-State Circuits Conference, pp 216–217
73. Song B, Na T, Kim J, Kim JP, Kang SH, Jung SO (2015) Latch offset cancellation sense amplifier for deep submicrometer STT-RAM. IEEE Trans. Circuits Syst. I: Reg. Papers 62(7):1776–1784
74. Na T, Kim J, Kim JP, Kang SH, Jung SO (2015) A double sensing-margin offset-canceling dual-stage sensing circuit for resistive nonvolatile memory. IEEE Trans Circuits Syst II: Exp Briefs 62(12):1109–1113
75. Jeong G, Cho W, Ahn S, Jeong H, Koh G, Hwang Y, Kim K (2003) A 0.24-μm 2.0-V 1T1MTJ 16-kb nonvolatile magnetoresistance RAM with self-reference sensing scheme. IEEE J. Solid-State Circuits (JSSC) 38(11):1906–1910
76. Motaman S, Ghosh S, Kulkarni JP (2015) A novel slope detection technique for robust STTRAM sensing. In: IEEE International Symposium on Low Power Electronics and Design (ISLPED), pp 7–12
77. Chen Y, Li H, Wang X, Zhu W, Xu W, Zhang T (2012) A 130 nm 1.2 V/3.3 V 16 Kb Spin-transfer torque random access memory with nondestructive self-reference sensing scheme. IEEE J Solid-State Circuits (JSSC) 47(2):560–573
78. Zeinali B, Madsen JK, Raghavan P, Moradi F. A novel nondestructive bit-line discharging scheme for deep submicrometer STT-RAM. IEEE Transactions on Emerging Topics in Computing
79. Toifl T, Menolfi C, Buchmann P, Kossel M, Morf T, Schmatz ML (2009) A 1.25–5 GHz clock generator with high-bandwidth supply-rejection using a regulated-replica regulator in 45-nm CMOS. IEEE J Solid-State Circuits (JSSC) 44(11):2901–2910
80. Phyu MW, Fu K, Goh WL, Yeo KS (2011) Power-efficient explicit-pulsed dual-edge triggered sense-amplifier flip-flops. IEEE Trans Very Large Scale Integr (VLSI) Syst 19(1):1–9

Chapter 2
Sensing of Resistive RAM

Qing Yang, Bonan Yan and Hai Li

2.1 Introduction

Resistive random-access memory (ReRAM) is a promising non-volatile memory with the configurability of resistance programmed by pulse voltage or current. ReRAM features a high integration density, long device endurance, fast access speed, and low read/write power, which give ReRAM an opportunity at the increasing gap between conventional volatile memory (e.g., DRAM) and non-volatile memory (e.g., hard disk drive). Beyond ReRAM's intrinsic role as an emerging memory device, its resistive nature gives a way to implement processing-in-memory (PIM) in a straight-forward strategy. While reading ReRAM cells, current accumulation by the Kirchhoff current formula can give an efficient solution of multiply-and-accumulate (MAC) operations.

In this chapter, we will firstly depict a whole view of the promising applications of ReRAM. Around these applications, the ReRAM sensing schemes domain the performances of memory applications and computation applications. In the following, this chapter will focus on two aspects in terms of two different application scenarios: sensing for ReRAM as memory and sensing for PIM using ReRAM.

Q. Yang · B. Yan · H. Li (✉)
Department of Electrical and Computer Engineering, Duke University, Rm 130 Hudson Hall, Durham, NC 27701, USA
e-mail: hai.li@duke.edu

Q. Yang
e-mail: qing.yang21@duke.edu

B. Yan
e-mail: bonan.yan@duke.edu

© Springer International Publishing AG, part of Springer Nature 2019
S. Ghosh (ed.), *Sensing of Non-Volatile Memory Demystified*,
https://doi.org/10.1007/978-3-319-97347-0_2

31

Table 2.1 Comparison between ReRAM with conventional memory technologies

Feature	SRAM	DRAM	NAND FLASH	ReRAM
Cell size (F^2)	123–140	6–8	4	4–12
Read time (ns)	~10	50–150	40–50	3–8
Write time (ns)	~10	50–150	~200	3–8
Endurance	–	–	10^5	10^{10}–10^{12}
Non-volatility	No	No	Yes	Yes

2.1.1 Advantages as Memory Cell and Data Processing Unit

There appears a speed gap between volatile memory and non-volatile memory. On one hand, with the rapid development of the semiconductor industry, the integrity density and speed of SRAM and DRAM are bringing higher data throughputs. On the other hand, the HDD is much slower which impedes the system performance. Flash memory is an appropriate choice to fill in the increasing gap. The weakness of Flash memory is the endurance which might cause a wear out problem, especially for enterprise users who expect a safe storage system.

The comparison between ReRAM and conventional memory technologies is shown in Table 2.1. Beyond the nature of non-volatility, ReRAM also features high integration density and fast access speed which are comparable or even better than SRAM and DRAM. Furthermore, the longer endurance of ReRAM compared to Flash memory gives us a better option for storage solution. With the advantages of high integration density, fast access speed, and long endurance, ReRAM gives us chances to have more substantial and reliable memory solutions in the computer memory hierarchy.

Besides acting as memories, ReRAM is also widely explored as processing unit in the field of processing-in-memory (PIM) [1–4]. Modern computers are heavily adopting the Von Neumann architecture as shown in Fig. 2.1a. In Von Neumann architecture, arithmetic unit and memory unit are separate. The control unit will fetch instructions and data from memory, and commonly, there is direct memory access (DMA) unit to help access data in memory directly. In memory-intensive tasks, e.g., neural networks, the data communication cost between the arithmetic unit and memory unit will domain the performance and power consumption.

As depicted in Fig. 2.1b, PIM structure enhanced with ReRAM manages to alleviate enormous data movements by combining data storage and computation. While reading the data programmed in ReRAM devices located at the cross-points of horizontal and vertical wires, it completes a multiplication operation through ReRAM's I-V characteristic. The bit-line current is the accumulation of multiplications on the same column. The reduction of data movements successfully saves time and power consumption . With the help of crossbar structure, ReRAM-based PIM has obtained

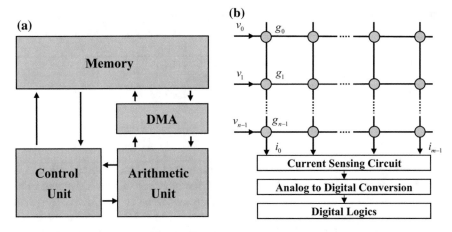

Fig. 2.1 (a) Von Neumann architecture; (b) typical PIM architecture using ReRAM crossbar

opportunities in various applications, e.g., deep learning, graph computing, and spiking neuron modeling, where it gets higher outcomes compared to general CPU and GPU platforms.

2.1.2 Sensing Scheme Is Important for System Performance

Considering the promising gains from ReRAM devices in memory design and PIM design, the key to obtain the high performance is the sensing scheme used in dedicated applications. In memory designs, one ReRAM cell is usually programmed with 1–2 bits resolution, and each ReRAM is sensed individually for memory reading. In contrast, under the scenario of PIM design, the sensing circuit shall read out the accumulative current from a ReRAM series with an acceptable accuracy for further arithmetic logic. Considering different application requirements, the determinant perspective of sensing for ReRAM as a memory cell is to have fast read/write circuit design to guarantee the fast response property of ReRAM, while the sensing accuracy is affected by the different resistance between different ReRAM states. In contrast, under the scenario of PIM design, the key problem turns out to be making trade-offs between speed and sensing accuracy. For example, in the neural network designs based on ReRAM, the data precision of interlayer is commonly expected be more than 8 bits.

The sensing scheme dominates the system performance, especially in PIM designs. Here we compare two commonly used sensing schemes in PIM DNN accelerators. In conventional ReRAM-based accelerator designs, digital-to-analog converter (DAC) and analog-to-digital converter (ADC) are commonly adopted as the peripheral circuitries of ReRAM crossbar for input data and computation results,

Table 2.2 Comparison between two different sensing solutions

Designs	Technology (nm)	Area (μm^2)	Static power (mW)	Active power (mW)
AD/DA design	32	1-bit DAC: 0.17 8-bit ADC: 1200	NA	2
Spiking design	150	1-bit pulse generator: 9.4 IFC: 396	0.0083	1.69

which brings big costs in power consumption and chip area. Another solution is to use the brain-inspired method to implement the interface of ReRAM-based PIM. Firstly, input data of crossbar and inter-core communication are modulated by spikes generated by a specific decoder, which helps get rid of DAC. Secondly, integrate-and-fire circuit (IFC) is utilized to complete the conversion from sensing current to digital value which helps get rid of ADC. We compare these two solutions in Table 2.2, where the measurements of AD/DA design are from [2], and the spiking-based design is simulated through Cadence Specter. From Table 2.2, although the spiking-based solution is implemented in 150 nm process other than a much more advanced 32 nm process in [5, 6], the combination of spike modulation and IFC can save more than 15% power with only 34% chip area cost. This advantage of spiking solution mainly comes from less demands on operational amplifiers that are heavily needed by DAC and ADC designs. It is also worth noting that the static power of spiking solution can get down to 0.0083 mW.

2.2 Sensing for Memory

2.2.1 ReRAM Characteristics as a Memory Cell

Predicted by Chua [7], memristor is the fourth fundamental circuit element defining the relation between magnetic flux (Φ) and electrical charge (q) as $d\Phi = M \cdot dq$. ReRAM is a typical kind of memristor. The resistance state of a ReRAM can be programmed by applying current or voltage. In 2008, HP Laboratory first reported their discovery of a nanoscale memristor based on TiO_2 thin-film devices [8]. Since then, many resistive materials and structures were found or rediscovered.

Figure 2.2 depicts an ion-migration filament model of metal oxide ReRAM device [9]. A metal oxide layer is sandwiched between two metal electrodes. During the reset process, the memristor switches from the low-resistance state (LRS) to the high-resistance state (HRS). The oxygen ions migrate from the electrode/oxide interface and recombine with the oxygen vacancies. A partially ruptured conductive filament region with high resistance per unit length (R_{off}) and a conductive filament region with low resistance per unit length (R_{on}) are formed; during the set process, the

Fig. 2.2 Metal oxide
ReRAM device

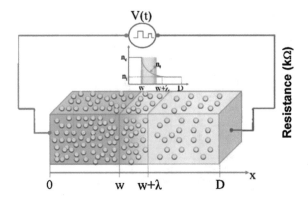

memristor switches from HRS to LRS as the ruptured conductive filament region shrinks. The resistance of many types of ReRAMs can be programmed to an arbitrary value (e.g., multi-level states) by applying a current or voltage with proper pulse width or magnitude. In many cases, the ReRAM resistance changes only when the applied voltage is above a threshold, e.g., V_{wrth}. With the feasibility of ReRAM that can be programmed onto different resistance levels, one ReRAM cell can be utilized to store bit information efficiently. In the read phase, a small read voltage lower than V_{wrth} will be applied to sense data out without disturbing the stored resistive states.

2.2.2   Read Circuitry

There are two common ReRAM sensing circuitry design schemes: voltage sensing and current sensing. We will use binary ReRAM devices as the example to illustrate the sensing strategy. Sensing for multi-level ReRAM cells can be easily extended by implementing multi-level reference cells.

Figure 2.3a shows a conventional ReRAM crossbar structure, where bit-line (BL) and source-line (BL) can be interchanged. The ReRAM cell is selected through combinations of word-line (WL), SL, and BL. In voltage-sensing scheme, before reading a ReRAM cell, the corresponding BL will be pre-charged to V_{PRE} as shown in Fig. 2.3b. Then, the WL will be set to select the row where the cell exists. The SL connecting to the other side of the cell will be grounded to start the BL discharging process. As depicted in Fig. 2.3b, the BL voltage drops with time. After an interval of time, the cell with high resistance will have a higher voltage level compared to the cell with low resistance level. The voltage gap provides us with a detection window. Through setting a reference voltage V_{REF} at the middle of voltage-sensing gap, we can easily determine the resistance states using a high-speed comparator.

It is worth noting that there exists a wide distribution of ReRAM resistance variations in both high-resistance state (HRS) and low-resistance state (LRS), which brings a wide distribution of discharging curves as in Fig. 2.3b. Before determining

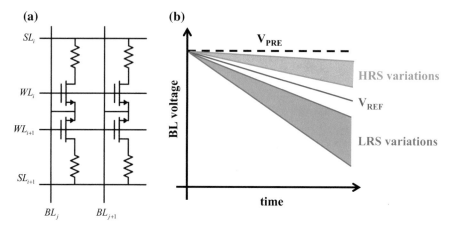

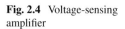

Fig. 2.3 Voltage-sensing scheme (a) ReRAM crossbar structure, (b) BL voltage swing

Fig. 2.4 Voltage-sensing amplifier

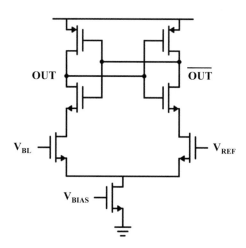

the V_{REF} for voltage sensing, the ReRAM device process should be explored to get precise resistance distribution. In practical ReRAM chip design, it is common to adopt auxiliary ReRAM array which is pre-programmed to generate on-chip voltage reference.

The high-speed voltage comparator is depicted in Fig. 2.4. The BL voltage V_{BL} after discharging and voltage reference V_{REF} will be fed into the input of differential pair. The cross-pulled inverter structure can have a high-speed response. For a voltage comparator that needs a bigger input voltage difference margin, the BL discharging time can be prolonged at the cost of slower read access speed.

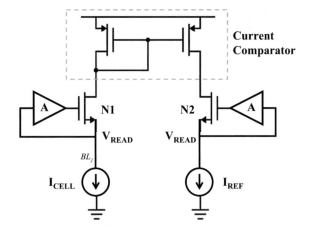

Fig. 2.5 Current-sensing scheme

Compared to voltage-sensing scheme, current sensing is also widely adopted. For circumstances with long BL arrays and small ReRAM cell currents, current-sensing scheme is usually better than the voltage-sensing scheme because current-sensing amplifier is more immune to BL noises.

Figure 2.5 shows the typical current-sensing scheme. With the help of a high-gain amplifier, the current-sensing scheme will clamp a small read voltage V_{READ} for the selected ReRAM cell at the source end of transistors N1. Meanwhile, the same V_{READ} is set at the source end of transistor N2 for pre-programmed reference ReRAM cells to generate the reference current I_{REF}. When one ReRAM cell is selected by WL and its SL connection is tied down to ground, the cell current I_{CELL} will be sensed out from BL. The comparison between I_{CELL} and I_{REF} will be completed in the simple current comparator to determine the ReRAM's resistance state. To maximize the current-sensing margin, I_{REF} should be set in the middle point of the read currents of high-resistance cell and low-resistance cell.

Besides the basic principles we have discussed ReRAM sensing schemes, many advanced techniques have been exploited to improve the sensing accuracy and speed. For voltage-sensing scheme, a swing-sample-and-couple (SSC) sense amplifier is proposed in [10]. SSC utilizes switch combination around a sensing capacitor to enlarge the voltage-sensing margin by about two times. On the current-sensing scheme side, many efforts are paid to generate the accurate reference current and the stable read voltage on BL. In [11–13], a parallel–series reference cell (PSRC) scheme is implemented to narrow the reference current distribution caused by ReRAM process variations. Instead of only using two ReRAM devices with high resistance and low resistance, respectively, PSRC utilizes several serial-connected and parallel-connected ReRAM reference blocks to generate the reference current. Furthermore, a process-temperature-aware dynamic BL-bias circuit is also considered in [12], which aims to achieve a process-aware read voltage on BL to overcome read disturbance due to different process corners.

Fig. 2.6 Typical write circuitry

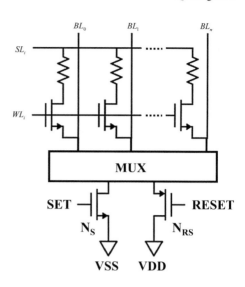

2.2.3 Write Circuitry

In the ReRAM writing phase, the write voltage will be higher than the device programming threshold V_{wrth}. As depicted in Fig. 2.6, in the write circuitry, every n BLs share one write driver. Write operations include SET and RESET for different programming directions. For SET operation, WL_i and SL_i are set high, and N_S is opened by SET signal to increase the ReRAM conductance. For RESET, WL_i is set high and SL_i is grounded; N_{RS} is opened to generate inverse programming current versus set operation, which decreases the ReRAM conductance. The column MUX works to control which column should be modified. To realize multi-level programming ability, N_S and N_{RS} are controlled by a pulse generator whose output pulse width is adaptable.

For ReRAM array, write operation costs more power than read operation. In the set process (i.e., ReRAM transfers from high resistance to low resistance), there will be a big rise in the write current; especially when several ReRAM cells on the same WL are set simultaneously. In [10], a self-boost-write-termination (SBWT) scheme is used to automatically cut-off the write current of ReRAM cells at once they have been set, which can improve write efficiency significantly. In addition, frequent writing operations will harm the endurance of ReRAM devices. In [14], they propose a configurable ramped voltage write scheme to address the endurance challenge in ReRAM, where set voltage is configured in one proper step, and reset voltage range is configured in four steps.

Fig. 2.7 Scenario of
matrix–vector multiplication

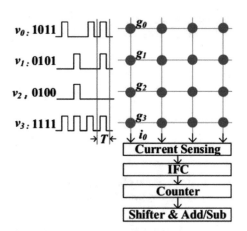

2.3 Sensing for Computing

2.3.1 Neuromorphic Computing

Figure 2.7 depicts the scenario of matrix–vector multiplication where 4-spike
sequence represents the input data. Spikes at different positions in the sequence can
have different weights. Hence, the input vector $\vec{V}$ can be a 4-bit binary data rather
than a 1-bit data (e.g., as IBM TrueNorth chip). The current along each bit-line (BL)
shows the dot production of vector $\vec{V}$ and $\vec{G}$, or $i_0 = \sum_{i=0}^{n} v_i \cdot g_i$, where v_i is the
pulse voltage applied on the source-line SL_i, and g_i ($i = 1...4$) are the conductance
of the ReRAM cells along the same BL. In the first spike period T0, a counter will
count the pulse number from the IFC and buffer the result in an adder; In the second
spike period T1, the obtained pulse number will shift 1 bit to the left (i.e., multiply
by 2) and will be added into the result from the first spike period. As we always
shift the pulse number obtained by $i-1$ bits to the left in the ith spike period, the
fixed-point computation output of the circuit is temporally encoded.

It is worth noting that v_i and g_i can only be positive values. Hence, for a $\vec{V}$ with
negative values, we can use 2's complementary representation to present the data.
For an n-bit input data, for example, the most significant bit has the weight of -2^n.
Thus, in the nth spike period, the pulse number obtained from the IFC will undergo a
shift-and-subtraction instead of a shift-and-addition. For a $\vec{G}$ with negative values, we
map it onto two BLs representing the positive value part $\vec{G}_+$ and the negative part $\vec{G}_-$,
respectively. The dot production result can be then calculated as the subtraction of
two BL currents $i_0 = i_{(0,+)} - i_{(0,-)} = \vec{V}(\vec{G}_+ - \vec{G}_-)$. As a result, the ReRAM-based
computing core can perform the signed fixed-point dot production, which can be
easily utilized to accelerate matrix–vector multiplication, $\vec{I}_{(m \times 1)} = \vec{G}_{(m \times n)} \cdot \vec{V}_{(n \times 1)}$.

Fig. 2.8 Small-signal model
for BL current conversion

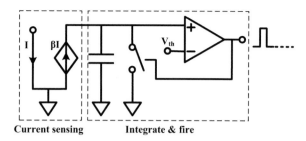

Current sensing   Integrate & fire

Note that the signal obtained from each BL is indeed the vector–vector multiplication
results of two vectors that are represented by the spike input and the ReRAM cells
on the BL.

The small-signal model for BL current conversion is depicted in Fig. 2.8. The
BL current is firstly buffered by current-sensing circuit shown before entering IFC's
capacitor, which will help stabilize the bit-line end voltage. Otherwise, turbulent
bit-line voltage would harm the computation accuracy, and the charge might sneak
reversely into the crossbar. Meanwhile, the current-sensing circuit can scale BL currents by tuning the current mirror at the output stage. The buffered BL current will
be available at the output stage and enter the following IFC. Initially, capacitor of
IFC is grounded by the discharging switch. With charge accumulation, the output of
comparator will turn high. The high output will be buffered by a delay-controlled
inverter chain before reaching the capacitor discharging switch. Then the accumu-
lated charges on capacitor will be discharged, leading the comparator output to turn
low again, which makes a pulse and so on. These generated pulses will be buffered at
the output stage. The last step of BL current conversion is to count the output pulses
of IFC in a unified time interval.

In the following, we will discuss the detailed designs of current amplifier and IFC
focusing on the sensing accuracy and stability discussion.

2.3.2 Current Amplifier

Figure 2.9 depicts the circuit diagram of current amplifier adopted in neuromorphic
computing for BL current sensing. Two high-gain operational amplifiers (Opamp) are
utilized. The open-loop gain is usually required to be higher than 60 dB, which can
be achieved by a single-stage folded-cascade structure. $Opamp_1$ is used to stabilize
the BL end voltage as V_{REF}, which prevents reverse sneaking path. For V_{REF} is set
in the range from 100 to 200 mV to allow a sufficient voltage swing on ReRAM
devices, the transistors N1 and N2 shall have a large channel width to have enough
current-sensing capability. $Opamp_2$ is used to keep the accuracy of current mirror
consisting of N1 and N2. The input current is explicitly buffered into the output

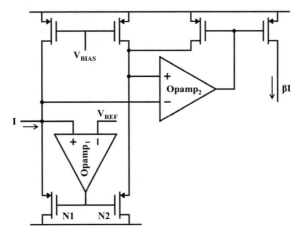

Fig. 2.9 Current amplifier diagram

current mirror stage. With the configuration of current mirror ratio β, the current amplifier can be adapted for various BL current ranges.

In the current amplifier design as shown in Fig. 2.9, there exists a loop consisting of Opamp$_1$ and N$_1$, which introduces oscillation risks. The open-loop analysis is needed to guarantee the stability. After unlooping as depicted in Fig. 2.10, there is a two-stage amplifier, where the first stage is the Opamp$_1$, and the second stage comprises the wide transistor N1. C_c is introduced by the parasitic capacitance C_{gd} of transistor N1, while C_{n1} is introduced by the parasitic capacitance C_{gs} of transistors N1 and N2. The load capacitance C_L is determined by the input differential pair of Opamp$_1$.

There are two poles and one positive zero in the two-stage amplifier. The dominant pole is at the output of the first stage determined by C_c, and the non-dominant pole is located at the output of the second stage determined by C_L. To prevent the circuit from oscillation, both dominant pole f_{nd} and positive zero f_z shall be kept far away from the dominant pole f_d. Usually, both f_{nd} and f_z will be placed at three times away from gain–bandwidth of the two-stage amplifier. The GBW is controlled by the transconductance g_{m1} of the first stage:

$$\text{GBW} = \frac{g_{m1}}{2\pi C_c}, \tag{2.1}$$

while f_{nd} and f_z are:

$$\begin{cases} f_{nd} = \dfrac{g_{m2}}{2\pi C_L} \cdot \dfrac{1}{1 + \frac{C_{n1}}{C_c}}, \\[3mm] f_z = \dfrac{g_{m2}}{2\pi C_c} \end{cases} \tag{2.2}$$

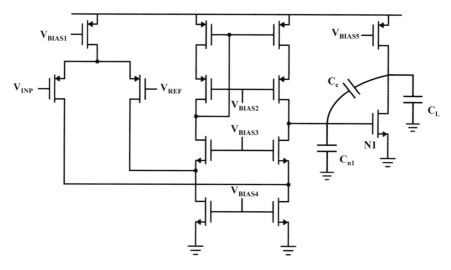

Fig. 2.10 Open-loop analysis of current amplifier

where g_{m2} is the transconductance of the second stage. To guarantee the stability, it should obey:

$$\begin{cases} f_{nd} > 3 \cdot \text{GBW} \\ f_z > 3 \cdot \text{GBW} \end{cases} \quad (2.3)$$

Solving Eq. (2.3), we can get the relationship between g_{m1} and g_{m2}, which guides the settings of bias currents for Opamp$_1$ and input transistor N1. Another way to help the stability is to add a compensation capacitor parallel with C_c that is to enlarge C_c. From Eq. (2.1), larger C_c leads GBW to be smaller, pushing f_{nd} much further away from the dominant pole.

2.3.3 IFC

IFC design features high speed and low power consumption. The structure is depicted in Fig. 2.11. The current buffered from current amplifier enters the input capacitor to rise the voltage to exceed V_{th}. Then the input differential pair together with the following cascaded inverters generates a high voltage at the gate of the discharging transistor N_{DISC}, which in turn enables N_{DISC}. Consequently, the voltage on input capacitor decreases quickly and eventually reverts the input differential pair. As such, the firing of one output spike is completed and a new iteration of integrate-and-fire starts. The generated spikes are buffered using an inverter chain for the followed operations in digital domain.

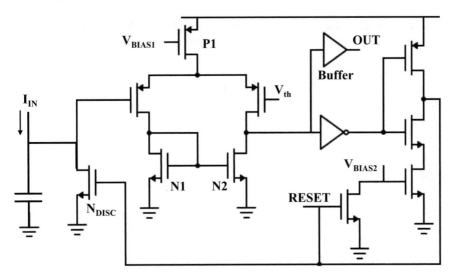

Fig. 2.11 IFC circuit diagram

To set V_{th}, there are two main considerations. First, in the initial condition, V_{th} should let all the transistors in the symmetrical operational amplifier work in the saturation region. V_{th} should be set in a boundary region, which gives transistors N1, N2, and P1 enough working margins for their saturation requirements. Second, to make sure N_{DISC} has an appropriate switching ability, the switching resistance should be much less than the output resistance of current amplifier. For N_{DISC} works in linear region, the switching resistance is:

$$R_{DISC} \approx \frac{1}{\mu_n C_{ox} \frac{W}{L}(V_{gs} - V_{thn})}, \tag{2.4}$$

where μ_n is the charge-carrier effective mobility, C_{ox} is the gate oxide capacitance per unit area, V_{gs} is the gate-to-source bias voltage, and V_{thn} is the threshold voltage of N_{DISC}. Deriving the switch resistance from Eq. (2.4), we can get the voltage drop ΔV_{DISC} of transistor N_{DISC} while discharging, and thus, we should set the threshold voltage V_{th} of IFC to be higher than ΔV_{DISC}. Otherwise, the inversion process of the input differential pair will get stuck.

2.4   Conclusive Remarks

In this chapter, we talked about the promising applications of ReRAM in memory design and PIM design. The sensing strategies for ReRAM play a dominant role in measuring the system performance. In the memory design, high-speed voltage/current-sensing scheme is required to handle high data throughputs. To circumvent device variations, reference voltage/current used in sensing comparators is generated by on-chip dummy ReRAM cells. PIM design based on ReRAM has different requirements on sensing strategy. The ReRAM sensing circuitry should meet two requirements: high power efficiency and appropriate data representation accuracy. Current amplifier and IFC are thoroughly discussed, including circuit structures, stability analysis, and key reference voltage settings.

References

1. Song L, Qian X, Li H, Chen Y (2017) PipeLayer: a pipelined ReRAM-based accelerator for deep learning. In: High performance computer architecture (HPCA), 2017 IEEE International Symposium on, pp 541–552. IEEE
2. Shafiee A, Nag A, Muralimanohar N, Balasubramanian R, Strachan JP, Hu M, Williams RS, Srikumar V (2016) ISAAC: a convolutional neural network accelerator with in-situ analog arithmetic in crossbars. In: Proceedings of the 43rd international symposium on computer architecture, pp 14–26. IEEE Press
3. Chi P, Li S, Xu C, Zhang T, Zhao J, Liu Y, Wang Y, Xie Y (2016) Prime: a novel processing-in-memory architecture for neural network computation in ReRAM-based main memory. In: Proceedings of the 43rd international symposium on computer architecture, pp 27–39. IEEE Press
4. Song L, Zhuo Y, Qian X, Li H, Chen Y (2017) GraphR: accelerating graph processing using ReRAM. arXiv preprint arXiv:1708.06248
5. Liu C, Yan B, Yang C, Song L, Li Z, Liu B, Chen Y, Li H, Wu Q, Jiang H (2015) A spiking neuromorphic design with resistive crossbar. In: Design automation conference (DAC), 2015 52nd ACM/EDAC/IEEE, pp 1–6. IEEE
6. Jiang H, Zhu W, Luo F, Bai K, Liu C, Zhang X, Joshua Yang J, Xia Q, Chen Y, Wu Q (2016) Cyclical sensing integrate-and-fire circuit for memristor array based neuromorphic computing. In: circuits and systems (ISCAS), 2016 IEEE international symposium on, pp 930–933. IEEE
7. Chua L (1971) Memristor-the missing circuit element. IEEE Trans Circuit Theory 18(5):507–519
8. Strukov DB, Snider GS, Stewart DR, Williams RS (2008) The missing memristor found. Nature 453(7191):80–83
9. Zhang L, Chen Z, Joshua Yang J, Wysocki B, McDonald N, Chen Y (2013) A compact modeling of TiO2-TiO2–x memristor. Appl Phys Lett 102(15):153503
10. Chang M-F, Wu J-J, Chien T-F, Liu Y-C, Yang T-C, Shen W-C, King Y-C et al (2015) Low VDD$_{min}$ swing-sample-and-couple sense amplifier and energy-efficient self-boost-write-termination scheme for embedded ReRAM macros against resistance and switch-time variations. IEEE J Solid-State Circuits 50(11):2786–2795
11. Lo C-P, Lin W-Z, Lin W-Y, Lin H-T, Yang T-H, Chiang Y-N, King Y-C et al (2017) Embedded 2 Mb ReRAM macro with 2.6 ns read access time using dynamic-trip-point-mismatch sampling current-mode sense amplifier for IoE applications. In: VLSI circuits, 2017 symposium on, pp C164–C165. IEEE
12. Chang Meng-Fan, Sheu Shyh-Shyuan, Lin Ku-Feng, Che-Wei Wu, Kuo Chia-Chen, Chiu Pi-Feng, Yang Yih-Shan et al (2013) A high-speed 7.2-ns read-write random access 4-Mb embedded resistive RAM (ReRAM) macro using process-variation-tolerant current-mode read schemes. IEEE J Solid-State Circuits 48(3):878–891

13. Sheu S-S, Chang M-F, Lin K-F, Wu C-W, Chen Y-S, Chiu P-F, Kuo C-C et al (2011) A 4 Mb embedded SLC resistive-RAM macro with 7.2 ns read-write random-access time and 160 ns MLC-access capability. In: Solid-state circuits conference digest of technical papers (ISSCC), 2011, IEEE International, pp 200–202. IEEE
14. Han X, Jia Q, Sun H, Wang L, Wu H, Cai Y, Zhang F et al (2017) A 0.13 μm 64 Mb HfO x ReRAM using configurable ramped voltage write and low read-disturb sensing techniques for reliability improvement. In: Custom integrated circuits conference (CICC), 2017 IEEE, pp 1–4. IEEE

Chapter 3
Sensing in Ferroelectric Memories and Flip-Flops

Ahmedullah Aziz, Sandeep Krishna Thirumala, Danni Wang, Sumitha George, Xueqing Li, Suman Datta, Vijaykrishnan Narayanan and Sumeet Kumar Gupta

3.1 Introduction

While non-volatile storage in the form of hard drives, solid-state drives, FLASH memories are ubiquitous, introducing non-volatility closer to the processor has been a subject of continual quest [1, 2]. Such an exploration has been triggered by the emergence of new applications as well as by a sustained improvement in the processing capability of the digital systems, which demands high-capacity memories. The advent of mobile systems demands computing with low power budgets, which, among other techniques, entails reducing or if possible, eliminating the power consumption during the standby state. In that context, non-volatility in caches can be very attractive as it enables complete shutdown of the power supply (V_{DD}) during the standby state (as opposed to a mere V_{DD} reduction as in conventional processors [3]). Consequently, the standby power can be zero, thereby significantly enhancing the battery life of such systems. At the same time, most non-volatile technologies offer a much denser solution compared to conventional SRAMs [1], leading to the possibilities of high-capacity caches. Another class of compute platform that has garnered an immense attention is energy autonomous systems [4] that harvest energy from the environment and therefore do not require a battery to operate. Such platforms can be extremely useful for healthcare, defense, and other applications, as they eliminate the need for battery recharging and significantly reduce the weight of the

A. Aziz · S. K. Thirumala · S. K. Gupta (✉)
School of Electrical and Computer Engineering, Purdue University, 465 Northwestern Avenue, West Lafayette, IN 47907, USA
e-mail: guptask@purdue.edu

D. Wang · S. George · X. Li · V. Narayanan · S. K. Gupta
School of Electrical Engineering and Computer Science, Pennsylvania State University, State College, PA, USA

S. Datta
Department of Electrical Engineering, University of Notre Dame, South Bend, IN, USA

© Springer International Publishing AG, part of Springer Nature 2019
S. Ghosh (ed.), *Sensing of Non-Volatile Memory Demystified*,
https://doi.org/10.1007/978-3-319-97347-0_3

system. However, since the energy derived from the environment may be unreliable and intermittent [2, 4], the processors must have features to back up the system state in the event of a power failure. From that perspective as well, non-volatile caches can be very useful. An even better design would be if non-volatility can be introduced very close to the compute elements, for instance, in the flip-flops. If that can be achieved efficiently, the energy associated with data backup and restore would be much reduced due to the elimination of large data traffic between processor and caches. Hence, exploration of technologies that can enable the design of non-volatile memories and logic (while meeting the power-performance-density requirements) has been and is expected to remain a subject of active research [1–8].

To that effect, several non-volatile memory technologies are being explored with techniques at different levels of design abstraction being developed to bring the characteristics of the respective technologies close to those of an ideal universal memory [9]. Examples include spin-based memories [10], resistive memories [11], phase change memories [12], ferroelectric-based memories [13], and others. Each of these technologies has their own advantages and issues; and so far, none of them have been able to meet all the targets. However, a large body of research [14–18] is being directed toward mitigating the limitations of these technologies with several promising approaches emerging that have a large potential to enable energy-efficient, robust, dense memory modules.

Among these technologies, ferroelectric (FE)-based memories (FERAMs) have been explored with great interest as they offer a high-density, high-endurance solution. However, destructive read in FE capacitors [13] and process incompatibility of some ferroelectric materials such as lead zirconium titanate (PZT) [19] have been the major design bottlenecks. Recent demonstration of ferroelectricity in hafnium zirconium oxide (HZO) [20], which is highly compatible with the CMOS process, has re-ignited the interest in this technology. Moreover, with the possibilities of integrating FE in the gate stack of a transistor (leading to ferroelectric transistors or FEFETs) [21] has opened other design opportunities with this technology. Another unique opportunity enabled by FE materials is in the design of ferroelectric tunnel junctions (FTJs) [22], which employ ferroelectrics sandwiched between two nano-magnets to augment the distinguishability of magnetic tunnel junctions. With such rich possibilities of ferroelectrics, it becomes important to understand the design aspects and application-specific optimizations for these design possibilities. While a thorough treatment of the aforementioned designs has been performed in several works for individual technologies [23, 24] and as comparative studies [13, 25], this chapter aims at providing design perspectives in context of data sensing considering the different design options offered by FE capacitors and FEFETs. FTJs require a separate treatment and are out of the scope of this chapter. The discussion on the sensing implications for FE-based capacitors and transistors will be presented considering non-volatile memories (NVM) as well as non-volatile logic (NVL).

This chapter is organized as follows. We will start by providing an overview of the FE capacitors and FEFETs in Sect. 3.2, discussing the relevant characteristics to lay the groundwork for the subsequent sections. In that section, we will also discuss the existing circuit topologies of ferroelectric memories and non-volatile logic.

In Sect. 3.3, we will review various read/sensing schemes for FE-based memories. This will be followed by the discussion on sensing techniques in non-volatile logic (specifically non-volatile flip-flops) in Sect. 3.4. Section 3.5 will summarize this chapter.

3.2   Ferroelectric-Based Memories and Logic

Ferroelectric materials have exotic polarization-voltage (P-V) characteristics unlike regular dielectrics. The landscape of the hysteretic P-V characteristics (discussed later) enables the ferroelectric materials to be able to hold on to either positive or negative polarization even in the absence of any applied voltage. This property of ferroelectric materials has made them promising for memory applications. The ferroelectric material, in form of a capacitor can serve as a non-volatile memory. In addition, a transistor augmented with ferroelectric materials (ferroelectric transistor) can also be designed to yield non-volatile storage. A discussion on these is as follows.

3.2.1   Ferroelectric Capacitor

A ferroelectric capacitor is a metal-FE (insulator)–metal structure which behaves as a 'non-volatile' capacitor. Unlike its conventional counterparts (say dielectric or electrolytic capacitors), a FE capacitor possesses nonzero (positive or negative) charge even when the effective voltage (V_{FE}) across it is set to zero. This distinct property of the FE material arises due to its ability to retain its polarization. To better comprehend, let us consider a typical polarization-voltage (P-V) characteristics of a ferroelectric capacitor (Fig. 3.1). As can be clearly seen from this figure, the FE exhibits nonzero polarization at $V_{FE} = 0$ V. This inherent polarization in the FE is called the *remnant polarization* (P_R). To explain the dynamics of polarization switching, let us first assume that the FE capacitor has a negative remnant polarization ($-P_R$). Now, if V_{FE} is increased, the magnitude of the polarization decreases toward zero (state A→B in Fig. 3.1). Eventually, the polarization gets nullified at a certain positive V_{FE} which is denoted as the *coercive voltage* (V_C). For $V_{FE} > V_C$, polarization eventually switches from negative to positive (B→C). At this stage, even if the voltage across the FE is revoked ($V_{FE} = 0$ V), the polarization is retained at the positive remnant polarization ($+P_R$) level. To nullify this polarization completely, a voltage equal to the negative of coercive voltage ($-V_C$) needs to be applied. Similarly, the opposite polarization switching dynamics (positive→negative) can be described considering an initial positive remnant polarization in the FE. In this case, we need to apply $V_{FE} < -|V_C|$ to switch polarization type from positive to negative (state X→Y→Z). Important point to note is that the negative→positive or positive→negative polarization switching only occurs when the $V_{FE} > +V_C$ and $V_{FE} < -V_C$, respectively.

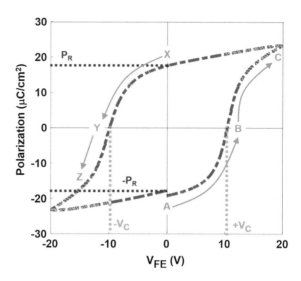

Fig. 3.1 Typical polarization versus voltage characteristics of a ferroelectric material

To understand the read mechanism for FE capacitors, let us note that the FE capacitor (C_{FE}) shows distinct high-capacitance regions near the positive/negative coercive fields and low-capacitance regions otherwise (Fig. 3.2). Consider that the FE is storing positive polarization. On applying a positive voltage across the FE ($V_{FE} > 0$), the entire trajectory of the FE polarization on the P-V loop will be in the low-capacitance region corresponding to $P > 0$ (Fig. 3.2). Thus, as the voltage across the FE increases, the current flowing through the FE ($I_{FE} = C_{FE}\,dV_{FE}/dt$) will be small. On the other hand, if the FE is storing negative polarization, application of sufficiently large positive V_{FE} will yield a trajectory from the low capacitance ($P < 0$) through high-capacitance (P switching from positive to negative) to low-capacitance region ($P > 0$). Thus the average capacitance increases leading to a large current. This difference in current is sensed to obtain the polarity of the stored polarization [25]. It is noteworthy that the read operation for FE capacitors is destructive; i.e., after sensing, the negative polarization switches to a positive one. Therefore, depending on the application requirements (as we will discuss subsequently), the sense operation may need to be followed by restoring the polarization in the FE capacitor.

Concisely, the FE capacitor may be set to hold positive or negative polarization with application of sufficiently high voltage of appropriate polarity. Having the capability of storing two types of charges makes it suitable to be used as a memory where the binary 1/0 bits are represented by the negative/positive polarization. Even after removal of the applied voltage, the polarization gets retained at $+P_R$ or $-P_R$ [25]. The remnant polarization (P_R) determines the retention of non-volatility in FE capacitor. On the other hand, the coercive voltage (V_C) acts as the threshold for programming/writing into the FE capacitor. V_C can be tuned with thickness of the FE (T_{FE}) (Fig. 3.3a) to make the switching in FE capacitor compatible with the operating voltages.

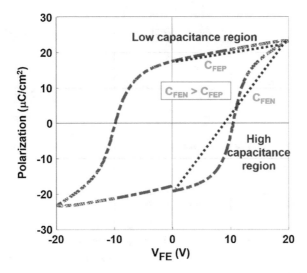

Fig. 3.2 Different levels of capacitance exhibited by ferroelectric capacitor across its polarization landscape

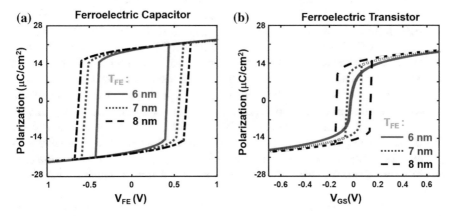

Fig. 3.3 **a** Polarization–voltage characteristics of FE capacitor with different thickness. The coercive voltage reduces with thickness. **b** Polarization–voltage characteristics of the gate stack of a ferroelectric transistor [23]

While using an FE capacitor directly as a memory element is well possible, the distinguishability of stored data and the robustness of sensing are major design challenges [25], as we will discuss subsequently. This concern can be well addressed and solved by using a ferroelectric transistor (FEFET) [23] which we describe next.

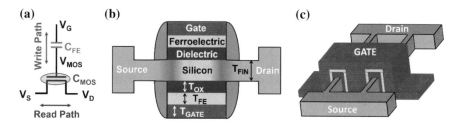

Fig. 3.4 Ferroelectric transistor (FEFET): **a** symbolic representation; **b** cross-sectional view; **c** corresponding 3D structure

3.2.2   Ferroelectric Transistor (FEFET)

A FEFET has a FE layer integrated in the gate stack (Fig. 3.4). The P-V response of an FEFET (in this case P-V_{GS} response, where V_{GS} is the gate-to-source voltage) shows hysteretic behavior similar to that of a standalone FE capacitor (Fig. 3.3b). However, the hysteresis window reduces for FEFET compared to an FE capacitor due to the effect of the capacitance of MOS structure of the FET and the associated depolarization fields [23]. That can be observed if we compare Fig. 3.3a with Fig. 3.3b. In fact, below a certain thickness ($T_{FE} = 7$ nm in Fig. 3.3b), the hysteretic property of the FEFET is completely lost and the FEFET supposedly enters the *steep switching* volatile mode [23]. Conceptually, this mode of FEFET was first proposed in 2008 by Salahuddin and Datta [26]. They envisioned driving the FE to take on an S-shaped trajectory while switching polarization. With proper design, the operation could be stabilized in the negative capacitance region of FE and eventually provide an amplified voltage in the gate stack of the transistor. This mode of operation for FEFET is currently being actively explored because of its promise to provide low-power logic operation. However, as this chapter only focuses on the non-volatile memory operation, we will refrain from going into details of the steep switching FEFETs.

Coming back to the context of non-volatile FEFET, we have seen from Fig. 3.3b, that with sufficiently large thickness of FE [23], we can achieve reasonable hysteresis in P-V_{GS} response of an FEFET. That means that even when the gate voltage (V_{GS}) is set to zero, the polarization in the FE will be retained. To switch the polarization to a positive value, sufficiently high V_{GS} needs to be applied such that the effective voltage across FE exceeds its coercive voltage. Similarly, a sufficiently negative V_{GS} can switch the polarization to the negative value. The bi-stability in the transfer characteristics of an FEFET has important implications in context of sensing. To explain that let us consider a p-FEFET. If this FET possesses a positive polarization in its gate stack, the electrostatics will not favor inversion of holes. Hence, in this scenario, the channel resistance of the FET will be high (therefore low drain current, I_D). On the other hand, a negative polarization store in gate stack makes it easier to form inversion layer in channel and thus reduces the channel resistance (therefore high I_D). As can be clearly seen from the I_D-V_{GS} characteristics of this p-FEFET

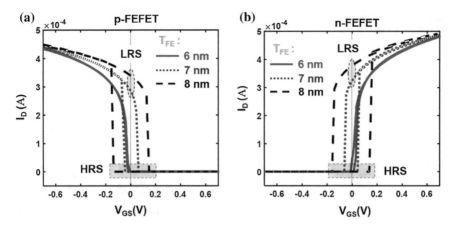

Fig. 3.5 Transfer characteristics of a **a** p-FEFET and **b** n-FEFET. The high-resistance state (HRS) and low-resistance state (LRS) have been marked for both versions of FEFETs

(Fig. 3.5a), the magnitude of current at $V_{GS} = 0$ V can exhibit orders of magnitude difference based on the stored polarization in the gate stack. Similarly, for an n-FEFET (Fig. 3.5b), positive polarization reduces the resistance of the FET (aids in channel inversion even at $V_{GS} = 0$), while negative polarization corresponds to the high-resistance state of the FEFET. Thus, the difference in polarization in gate stack gets translated to orders of magnitude difference in drain current as the gain of the transistor gets coupled with the dynamics of ferroelectrics. The polarization difference in the gate stack may be used to store logic 1/0, and the difference in drain current can be utilized to later sense the stored data, but with much more robustness compared to FE capacitors. We will talk about these in detail as we dive deep in this chapter. With the understanding of the operation of FE capacitors and FEFETs, let us now discuss some topologies of non-volatile memory and logic-based on these entities.

3.2.3 Ferroelectric Capacitor-Based Memory (FERAM) Topologies

Ferroelectric capacitors are coupled with transistor(s) to achieve a memory cell that can undergo read/write operation as desired. Such memory cells are termed as FER-AMs. The basic primitive of an FERAM has one access transistor to control the memory operation and one ferroelectric capacitor as storage element. However, in consideration of robustness in sensing and cell density, there can be variants of FER-AMs with multiple transistors and capacitors. We discuss these topologies below.

Fig. 3.6 Schematic of a
1T-1C FERAM

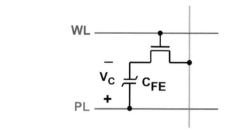

Fig. 3.7 Write mechanism
in a 1T-1C FERAM [25]

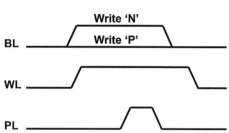

3.2.3.1 One Transistor-One Capacitor (1T-1C) Topology

The 1T-1C topology is the most basic primitive of FERAM. The schematic of 1T-1C FERAM is shown in Fig. 3.6. The FE capacitor is accessed in an array through an access transistor. The bit-line (BL) is shared along the column, while the word-line (WL) and plate-line (PL) are shared along the row. To perform the write operation (Fig. 3.7), BL voltage is driven to V_{DD} to switch polarization to a positive value while it is kept at 0 for programming negative polarization in the FE capacitor. WL is asserted and a pulse is applied at the PL [25]. During the time PL is at 0 V, cell with BL at V_{DD} has positive voltage across FE ($>V_C$), which switches the polarization to a positive value, When PL voltage switches to V_{DD}, the cells with BL at 0 V has negative voltage across FE ($<-V_C$), which switches it to a negative value. In the hold state, FE polarization is either at $+P_R$ or $-P_R$ (i.e., the remnant polarization—see Fig. 3.1). The read operation involves detecting the positive or negative value of the polarization. This is achieved by applying an appropriate voltage across the FE to sense the average capacitance for positive and negative polarization (see Fig. 3.2) and in turn the associated current change. In a memory array, this operation is achieved by applying an appropriate voltage on PL and sensing the voltage on BL. Note, this topology necessitates using reference capacitors for sensing the stored data. We will discuss details of its sensing method in Sect. 3.3.

3.2.3.2 Two Transistor-Two Capacitor (2T-2C) Topology

A 2T-2C topology is a self-referenced design, and hence, it averts the necessity for using reference capacitor. A 2T-2C FERAM cell (Fig. 3.8) has been explored in

Fig. 3.8 Schematic of a
2T-2C FERAM Cell [27]

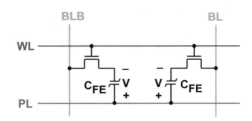

[27] which could effectively track the process variations. The two FE capacitances store complementary polarizations, and thus, the cell is differential in nature. This makes the 2T-2C implementation self-referenced, bypassing the design challenges associated with reference generation. Moreover, because of opposite swings in complementary bit-lines, the sense margin is double that of the conventional single-ended designs. The 2T-2C cell can employ BL voltage swing generated by employing the changing PL voltage to sense the logic state [25]. Alternatively, current source-based BL charging method can also be employed as in [28] to utilize the different time constants for BL charging to detect the logic state. For the latter technique, the authors in [28] propose further optimization by employing the p-transistors of the sense amplifier to also serve as the current sources for the complementary bit-lines. As the sense amplifier resolves the difference between the BL voltages, positive feedback dynamically reduces the current for branch corresponding to positive polarization. The results in [28] show that this technique leads to 40% improvement in read speed compared to the conventional scheme of using separate transistors for the current source. The 2T-2C design with or without such optimizations offers much better sense margins and read robustness compared to the conventional 1T-1C cells. However, it incurs more cost in terms of area as two transistors are required per cell.

3.2.3.3 Chain FERAM Topology

2T-2C cell may not be suitable for designs requiring ultra-high cell density. Such applications can benefits from the chain FERAM [29] illustrated in Fig. 3.9, which offers densities much higher than even the 1T-1C configuration. In this design, the unaccessed cells have their WLs at V_{DD}, due to which the respective access transistors bypass the corresponding FE capacitors. With very low voltage across the FE, the polarization is retained. The WL for the accessed cell is at 0, as a result of which the corresponding access transistor is OFF. Thus, the FE capacitance is connected to the BL and PL through a series of ON transistors and the read operation can be performed as described previously. The sensing in this design is more challenging as the BL voltage swing depends on the capacitance of the series transistors, in addition to the BL capacitance. However, since the contact of the series connected access transistors can be shared (as in NAND FLASH, from which this design draws inspiration [29]), very high densities can be achieved.

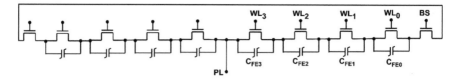

Fig. 3.9 Circuit diagram of two cell blocks in a chain FERAM architecture [29]

The fundamental mechanism for read and sensing FERAMs (capacitance difference between the two logic states) is dependent on switching the negative polarization (while retaining the positive polarization). Because of the destructive nature of the read operation, write-back operation is essential (discussed later), which deteriorates the power and performance the FERAMs. Now, we will discuss the ferroelectric transistor (FEFET)-based memories, which effectively counters this problem by enabling a non-destructive read and very high distinguishability between the two logic states.

3.2.4 Ferroelectric Transistor (FEFET)-Based Memory Topologies

Ferroelectric transistors (FEFETs) combine the non-volatility of the FE with the inherent gain in the transistor to achieve unique hysteretic transfer characteristics, as discussed before. By virtue of this, the polarization difference between the two logic states is automatically translated into high and low resistance of the FEFET. Furthermore, the inherent transistor gain leads to large ratio of the resistance corresponding to the two logic states (Fig. 3.5), which make it substantially easier to sense the logic state of the FEFET compared to FERAMs [30].

To design a memory cell based on FEFETs, multiple access transistors may be required [30] due to the separation of read and write paths in the FEFETs (Fig. 3.4a). To minimize the area overheads of multiple access transistors, different architectures are possible, each with its own sensing implications. In [30], a 2T design based on an FEFET and an access transistor in the write path is proposed (Fig. 3.10a). The read access transistor is avoided by connecting the FEFET between read word-line and read bit-line in a cross-point fashion. However, to simplify the biasing scheme, a separate read access transistor is required, leading to a 3T design (Fig. 3.10b), which isolates the accessed cell from the unaccessed cells. However, both 2T and 3T designs require negative write voltages on the write bit-line (WBL) for switching the FE polarization to a negative value [31]. To avoid the negative write voltages, a 4T design is required (Fig. 3.10c). This design enables sharing the read bit-line and source line (SL) along the column. Thus, to switch the polarization of FEFET to a positive value, positive write voltage is applied on the write bit-line and the source line is grounded. On the other hand, to induce negative polarization, positive write voltage can be applied on the source line while the write bit-line is driven to 0 V. The sensing implication of these topologies will be described in Sect. 3.3.

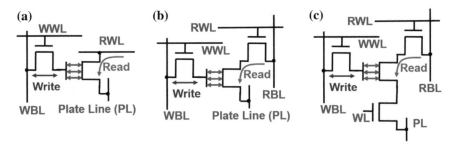

Fig. 3.10 Different topologies of FEFET-based memory. **a** 2T **b** 3T **c** 4T

In addition to the non-volatile memory applications, FE capacitors and FEFETs can be very beneficial for introducing non-volatility inside the logic [2]. Such non-volatile processor architectures can be useful for systems that need to back up the system state when the power goes off, either by design or accidentally. The former case is applicable to systems that are normally OFF or spend a large time in the standby state. To reduce the leakage power, it can be very useful to shut down the power supply during long standby periods. Non-volatility in logic can enable data backup so that the system remembers its previous state and resume computation more effectively once the power re-appears. The other class of systems, in which non-volatile logic can be very useful, is the energy autonomous platforms which operate with energy harvested from the environment. Since the energy source may be unreliable, power may go off intermittently. Non-volatility in logic enables backing-up the system state in the event of power failure. This, in turn, enables the processor to resume computation from where it had stopped (rather than to start the computation from the beginning as in standard volatile processors). In [4], it is shown that such architectures can lead to a significant increase in the forward computation progress when the energy source is unreliable. As mentioned before, data may be backed up in a non-volatile memory as well. However, that requires energy-intensive data flow between logic and memory. Therefore, introducing non-volatility in proximity to the logic can significantly improve the energy efficiency of data backup (during power OFF) and restore (when the power supply resumes). Since the system state typically resides in the flip-flops, the main focus of enabling non-volatile logic has been to introduce data backup/restore features in flip-flops. FE-based technologies have been explored to enable such circuits, and in the following subsections, we will provide an overview of the topologies for FE capacitors and FEFETs in context of the design of non-volatile flip-flops.

3.2.5 Ferroelectric Capacitor-Based Non-volatile Logic Topologies

Several designs have been proposed for non-volatile flip-flops with FE capacitors serving the role of non-volatile element in which the state of the system can be backed up. Broadly, the designs of non-volatile flip-flops can be categorized into two flavors. The first class of designs [32, 33] employs a separate backup/restore module, which can be coupled with a flip-flop to enable storing the logic state into the FE capacitor in the event of a power failure. When the power returns, the data stored in the capacitor are used to set or reset the flip-flop. In the second category, the FE capacitors are integrated inside the flip-flop architecture to introduce data backup and restore features [34]. The advantage of the first design is that any flip-flop topology optimized to meet the design targets can be coupled with the backup/restore module, with only minor modifications, if necessary. Moreover, the impact of introducing non-volatility on the speed and power during the normal operation can be minimized. The benefits of the second design are that with FE capacitors in close proximity to the storage nodes of the flip-flop (primarily those of the slave latch), backup/restore can be carried out with minimal control signals. A brief description of such topologies is as follows.

3.2.5.1 Ferroelectric Capacitor-Based Flip-Flop with Separate Backup/Restore Module

Figure 3.11 shows a non-volatile flip-flop with a separate back-up/restore module, which employs two FE capacitors [32]. The capacitors are connected to the complementary storage nodes of the slave latch of the flip-flop (Q and QB). On the application of a pulse at PL (similar to the write operation of 1T-1C FERAM—Fig. 3.7), the FE capacitor controlled by the storage node at '1' stores negative polarization when PL is at 0. On the other hand, the FE capacitors connected to the storage node at '0' switches its polarization to a positive value when PL is at V_{DD}.

Figure 3.12 shows a non-volatile flip-flop with four FE capacitors in the backup/restore module [33]. The outputs (Q and QB) of the flip-flop controls the FE polarization in a manner such that FE1 and FE3 store the same polarization (P_{13}), while the polarization of FE2 and FE4 (P_{24}) is opposite to that of P_{13}. This is achieved by applying a pulse at plate-lines PL1 and PL2, similar to the write operation of 1T-1C FERAM (Fig. 3.7). If Q = '1' (QB = '0'), FE1 experiences positive voltage when PL1 = V_{DD}; FE2 experiences negative voltage when PL2 = V_{DD}; FE3 experiences negative voltage when PL1 = 0; and FE4 experiences positive voltage when PL2 = 0. Therefore, for Q = '1', FE1 and FE4 store positive polarization, while FE2 and FE3 store negative polarization. Similarly, if Q = '0', FE1 and FE4 have negative polarization while FE2 and FE3 operate in the positive polarization states.

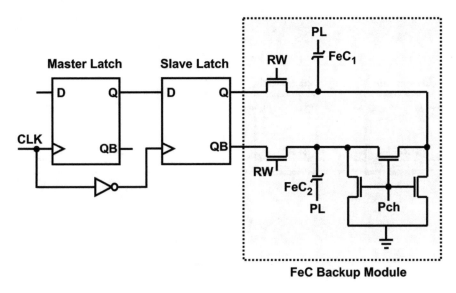

Fig. 3.11 Non-volatile flip-flop with 2 FE capacitors and separate backup/restore module [32]

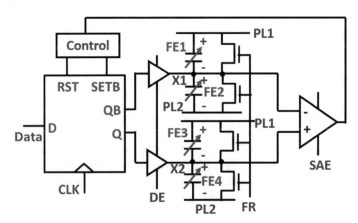

Fig. 3.12 Non-volatile flip-flop with 4 FE capacitors in the backup/restore module [33]

3.2.5.2 Ferroelectric Capacitor-Based Flip-Flop with Integrated Backup/Restore Module

Let us now take an example of the flip-flop with FE capacitors integrated in the flip-flop architecture. The block diagram is shown in Fig. 3.13 [34]. This design backs-up the data in a complementary fashion the two FE capacitances, similar to the design in Fig. 3.11. As the restore operation is closely coupled with sensing technique, we will discuss the sensing/restore mechanism later in Sect. 3.4.

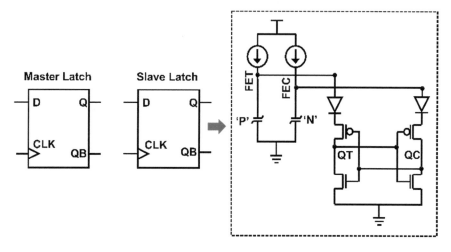

Fig. 3.13 Non-volatile flip-flop with integrated FE capacitors [34]

Ferroelectric transistors can also be used to obtain non-volatile logic functionality. In fact, they have some unique advantages over the ferroelectric capacitor-based designs. We discuss them in the following subsection.

3.2.6 Ferroelectric Transistor-Based Non-volatile Logic Topologies

Ferroelectric transistors (FEFETs), by virtue of the non-volatility built into the three terminal structure, offers very appealing design opportunities for introducing data backup and restore features in the flip-flops [23]. In other words, the FEFETs are inherently compatible to be integrated with the flip-flop (based on CMOS transistors) as they are yet another form a transistor (albeit now with non-volatile functionality). Moreover, the difference in the resistance of the FEFETs corresponding to the two logic states is enormous ($\sim 10^4$—see Fig. 3.5). This makes the sensing or restore operation very efficient. Similar to the FE capacitor-based non-volatile flip-flops, one has the option to either design a separate data backup/restore module or integrate FEFETs inside the flip-flop architecture.

3.2.6.1 Ferroelectric Transistor-Based Flip-Flop with Separate Backup/Restore Module

Figure 3.14 shows the flip-flop with a separate backup restore module employing one FEFET in conjunction with other transistors [23]. The unique feature of this

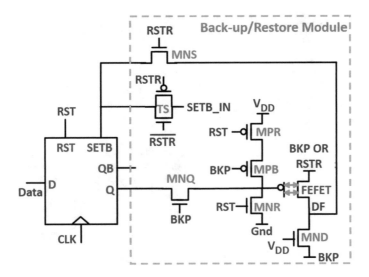

Fig. 3.14 Non-volatile flip-flop based on FEFET with separate backup/restore module [23]

flip-flop is that before the backup operation (i.e., during the normal operation), the bias conditions of FEFET as such that a positive polarization is stored in the p-type FEFET. Thus, FEFET is in the high-resistance state (see Fig. 3.5a). Therefore, during the backup operation, the polarization switching only needs to occur when flip-flop output $(Q) = 0$, i.e., when a negative polarization needs to be induced. As a result, the backup delay is incurred only in this case. Note, this technique avoids the use of a pulse during the backup operation because polarization needs to be switched only in one direction. (This is unlike FE capacitor-based design, which require PL pulse as discussed before, and therefore, backup delay can be as large as the half time period of the pulse + the delay associated with polarization switching). During backup, Q is connected to the gate of the FEFET and the source/drain is driven to V_{DD}. If $Q = 0$, the gate-to-source voltage of the FEFET becomes negative, as a result of which, negative polarization is induced.

3.2.6.2 Ferroelectric Transistor-Based Flip-Flop with Integrated Backup/Restore Module

Another design possibility is to integrate FEFETs inside the flip-flop architecture, as shown in Fig. 3.15 [35]. This design employs 2 FEFETs. The data backup occurs by controlling the gate of the FEFET with one of the storage nodes of the slave latch and its source/drain with the other storage node. Therefore, depending on the flip-flop state, the gate-to-source voltage is either negative or positive. Similarly, the other FEFET stores complementary polarization. Restore operation will be discussed along with sensing mechanism in Sect. 3.4.

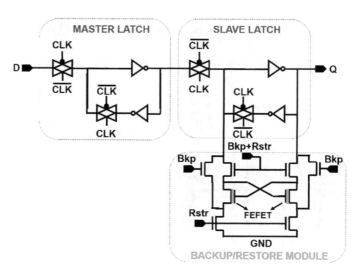

Fig. 3.15 Non-volatile flip-flop with FEFET integrated in the flip-flop [35]

With understanding of these circuit topologies and principle of operation, we will now move on to the prime focus of this chapter. We will now discuss the sensing techniques associated with ferroelectric memory and logic. The description of the following sections will be very closely coupled with the topologies and variants described so far. In Sect. 3.3, we will describe the sensing methods for ferroelectric memories. We will talk about different schemes for sensing along with peripheral design considerations. We will also correlate and compare the sensing process for ferroelectric capacitor and FEFET-based memories. In Sect. 3.4, we will provide a similar discussion for ferroelectric non-volatile logic.

3.3   Sensing Techniques for Ferroelectric Memories

In the previous section, we discussed how FE capacitors and FEFETs serve the function of non-volatile elements. Both the technologies utilize polarization retention in the FE to store the logic states. The write operation also entails a similar scheme in both, i.e., applying bipolar voltage across the FE capacitor or the gate-source of the FEFETs such that the magnitude of the voltage across FE is greater than the magnitude of its coercive voltage ($|V_C|$). However, sensing the data involves different mechanisms. For FE capacitors, the capacitance difference between the two states (and the consequent current difference) is employed to sense the logic value, while in FEFETs, the conductance difference for positive and negative polarizations is exploited to read the data. In this section, we will go into the details of the sensing techniques employed for FE capacitors (FERAMs) and FEFETs from the perspective of memory read.

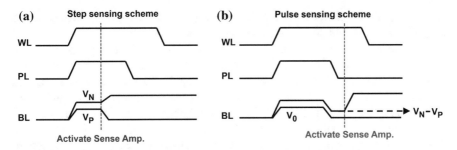

Fig. 3.16 Two typically used sensing schemes for 1T-1C FERAMs [25]. **a** Step sensing and **b** pulse sensing

3.3.1   Sensing Techniques for Ferroelectric Capacitor-Based Memories

As mentioned in Sect. 3.2, ferroelectric capacitors have completely different sense mechanisms compared to FEFET-based memories. Hence, in this subsection, we will restrict our discussion on different aspects of sensing for FE capacitor-based memories. We will first talk about the two primary methods for sensing in such memories. Then, we will direct attention toward the mechanism for designing peripherals.

3.3.1.1   Sensing Methods

Typically, two sensing methods are used in ferroelectric capacitor-based memories (illustrated in Fig. 3.16). The read operation is initiated by pre-discharging BL to 0 after which WL is asserted. In the first method (Fig. 3.16a), also called the step sensing approach [25], the PL voltage is increased, which either keeps the polarization of FE at the positive value or switches it from negative to positive (Fig. 3.1). In the former case, the current in the FE and therefore, the increase in the BL voltage (V_{BLP}) is small. But, the latter case results in a relatively larger increase in BL voltage (V_{BLN}), as can also be understood from the equations below.

$$V_{BLN} = [C_{FEN}/(C_{BL} + C_{FEN})]V_{DD} \tag{1a}$$

$$V_{BLP} = [C_{FEP}/(C_{BL} + C_{FEP})]V_{DD} \tag{1b}$$

Here, C_{FEN} and C_{FEP} are the average capacitances of the FE for negative and positive polarizations respectively, and C_{BL} is the bit-line capacitance. As discussed in the previous section, $C_{FEN} > C_{FEP}$ (see Fig. 3.2). The difference between the BL voltages ($V_{DIFF} = V_{BLN} - V_{BLP}$) is then utilized by a sense amplifier to discharge BL to 0 (if the original FE polarization was positive) or to charge it to V_{DD} (if the original FE polarization was negative). After this, PL is driven to 0 so that negative voltage appears across the FE for the latter case (with BL at V_{DD}) at which point the FE polarization is switched back to the original negative value.

In the second method (Fig. 3.16b), referred to as the pulse sensing approach [25], the sensing is performed by first increasing the PL voltage. Similar to step sensing, BL voltage increases to a larger extent when the initial polarization is negative compared to the positive polarization. Thus, the difference in the BL voltage = V_{DIFF}, as before. It may be recalled that after the process, the final polarization of the FE is positive, irrespective of the initial state of the FE. Before sensing, the PL voltage is decreased to zero, which leads to a decrease in the BL voltage. Since during PL voltage reduction, the polarization for the two cases remains positive, the decrease in BL voltage is the same. This means that the initial difference in the BL voltage for the two logic states is maintained. Hence, if the initial polarization was positive, BL voltage returns to 0 ($V_{BLP} = 0$); while if the initial polarization was negative, BL voltage $V_{BLN} = V_{DIFF}$. This difference is utilized by the sense amplifier to charge BL to V_{DD} or discharge it to 0 V. After full swing on BL is obtained, the voltage across FE is negative for BL voltage = V_{DD} (initial polarization < 0). Therefore, in this case, the FE polarization switches to a negative value and the initial logic state is restored (similar to step sensing).

The advantages of step sensing over pulse sensing are the following [25]. First, step sensing is faster since in the pulse sensing, the activation of sense amplifier is performed after waiting for PL voltage to go down. Second, the common mode BL voltage for step sensing ($=(V_{BLP} + V_{BLN})/2$) is larger than pulse sensing ($=V_{DIFF}/2$ $= (V_{BLN} - V_{BLP})/2$), which makes the biasing of sense amplifier more efficient. Third, FE capacitors storing positive polarization experience $0 \rightarrow V_{DD} \rightarrow 0$ cycle in step sensing (see Fig. 3.16a), while in pulse sensing, voltage cycle is $0 \rightarrow (V_{DD} - V_{BLP}) \rightarrow 0$. Full voltage cycle helps in maintaining the positive polarization in step sensing more effectively, thus improving the retention time. As a corollary, pulse sensing may require an extra pulse to reinforce positive polarization, which may decrease its performance. The advantage of pulse sensing is that the effect of process variations is mitigated. To understand this, let us recall that before activating the sense amplifier in pulse sensing, the polarization returns to its initial value for the case when the initial polarization was positive. On the other hand, in step sensing, the polarization sensed is larger than the stored value. This increase in polarization may be different for different cells due to process variations. As a result, the polarization and, therefore, the BL voltage sensed show a distribution for step sensing. However, in pulse sensing, since the polarization returns to its original value, the effect of process variations and, therefore, the distribution of sense BL voltage reduce. Since step sensing exhibits many more benefits compared to pulse sensing, the discussion in the subsequent subsections will be from the perspective of step sensing, unless stated otherwise.

So far, we have discussed the sensing of the positive and negative polarizations for the FE capacitor in context of the cell operation. Let us now present other design perspectives related to the read operation, particularly from the point of view of reference voltage generation and sense amplifiers.

3.3.1.2 Peripheral Circuits Design Consideration

Since the standard 1T-1C FERAMs are single ended, they require a reference voltage $(V_{REF}) = (V_{BLN} + V_{BLP})/2$, which is utilized by the sense amplifier to compare the BL voltage and drive it to V_{DD} or 0, as the case may be. The sensing techniques discussed previously rely on capacitance division between the FE capacitance and BL capacitance, as can be observed in Eq. (1a). Therefore, the sense margin is a strong function of C_{BL}, which implies that the sense margin and the reference voltage to be generated is dependent on the size of the array, non-linear capacitance of the access transistors and other factors. Moreover, the fatigue or retention loss due to repeated polarization switching and other effects such as imprint (preference of one polarization over the other because FE is in that state for a long time [36]) and relaxation (loss of remnant polarization if the FE is not switching for a long time immediately after continuous cycling [37]) can be different for different memory cells. Further, the reference voltage is typically generated employing reference FE capacitors (C_{FEREF}), which can experience different 'wear and tear' effects compared to the memory cells. Thus, it is very challenging to generate V_{REF} that is exactly mid-way the bit-line voltages corresponding to the two logic states. Figures 3.17, 3.18, 3.19, and 3.20 illustrate various options for voltage reference generation, an excellent comparison of some of which can be found in [25]. A brief categorized description for these approaches is as follows.

- **Single Reference Capacitor-Based Scheme**

The most straightforward approach is to use a single C_{FEREF} [38], which is programmed to store positive polarization (Fig. 3.17). C_{FEREF} is sized larger than the memory FE capacitances such that it is able to generate reference BL voltage that is between V_{BLN} and V_{BLP}. The reference capacitor needs to be biased appropriately such that it never experiences a negative voltage and positive-to-negative polarization switching. For this, additional transistors are used (not shown in the figures); the details of which can be found in [25]. Since C_{FEREF} does not switch its polarization, fatigue effects are minimized; however, other effects such as imprint degrade the difference between memory and reference bit-lines over time. The accuracy of this method to generate V_{REF} mid-way between V_{BLP} and V_{BLN} strongly depends on the size of C_{FEREF}. To improve the accuracy, dynamically adaptable C_{REF} can be used by employing incremental capacitances to change the overall C_{FEREF}. In [39], 128 FE capacitances controlled by 7-bit control vector are used to achieve a programmable reference. Alternatively, it has also been proposed that instead of C_{FEREF}, the plate-line voltage may be varied to obtain the target V_{REF} [40].

- **Dual Reference Capacitor-Based Scheme**

Another technique (Fig. 3.18), which is independent of the size of C_{FEREF}, is to use two reference capacitors storing positive and negative polarizations ($C_{FEREF,P}$ and $C_{FEREF,N}$, respectively) to generate the reference voltage [41]. In Fig. 3.18, the size of $C_{FEREF,P}$ and $C_{FEREF,N}$ is half the size of the memory capacitances (C_{FE}).

Fig. 3.17 Voltage reference
generation using a single
reference FE capacitor [38]

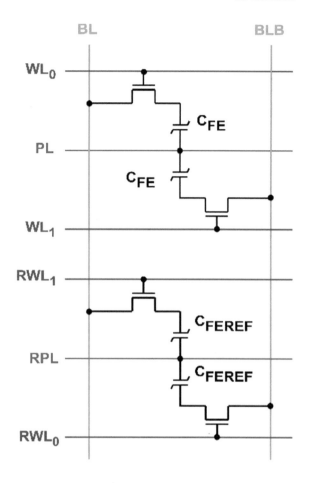

Since the BL voltage is affected by both the reference capacitances, $V_{REF} = (V_{BLN} + V_{BLP})/2$ is expected. However, BL capacitance also affects the reference voltage. The impact of BL capacitance on V_{REF} stems from the fact that $C_{BL}/C_{FEREFN,P} = 2\,C_{BL}/C_{FE}$ because of half-size of the reference capacitances. Thus, the bit-line voltage contributed by each capacitance is different from that in the memory cells. In general, BL capacitance increases V_{REF} compared to the ideal value [25]. Therefore, this method may not achieve the desired accuracy especially when C_{BL} is much larger than the FE capacitance. Furthermore, the reference capacitance $C_{FEREF,N}$ switches for every read operation and therefore is fatigued much more than the memory cells. An improvement over this method is to split the BL capacitance into two [25]. One half of BL interacts with $C_{FEREF,P}$ and the other half interacts with $C_{FEREF,N}$ (Fig. 3.19). This mitigates the issue with the previous technique by equalizing the ratio of BL and FE capacitances in the memory and reference cells. As a result, V_{REF} closer to the ideal value can be obtained. Once the two portions of the BL receive their respective charges from the reference FE capacitances, BLS is asserted

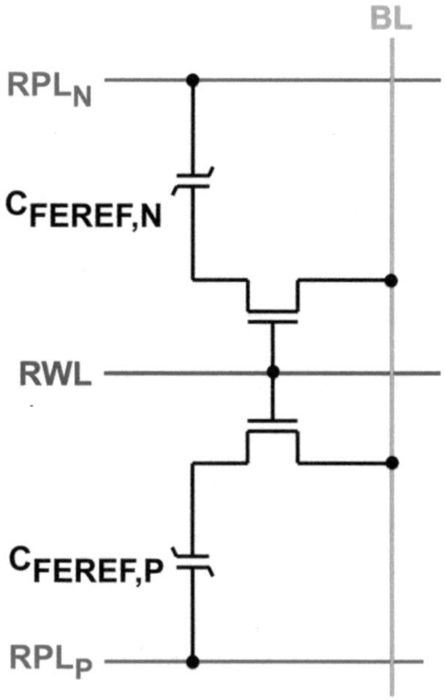

Fig. 3.18 Voltage reference generation using two reference FE capacitors [41]

Fig. 3.19 Voltage reference generation using two reference FE capacitors and split bit-line [25]

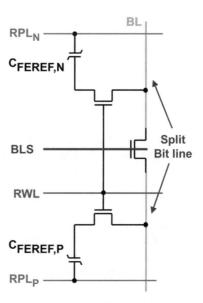

to enable charge sharing. Yet another improvement, as proposed in [42], is to use $C_{\text{FEREF,N}}$ and $C_{\text{FEREF,P}}$, which have the same size as the memory FE capacitor (C_{FE}), but are shared with two neighboring columns (Fig. 3.20). This averts mismatch in the BL voltages due to non-linearity of FE capacitances with respect to the size. One of the two neighboring BLs interacts with $C_{\text{FEREF,N}}$ and the other with $C_{\text{FEREF,P}}$ thus

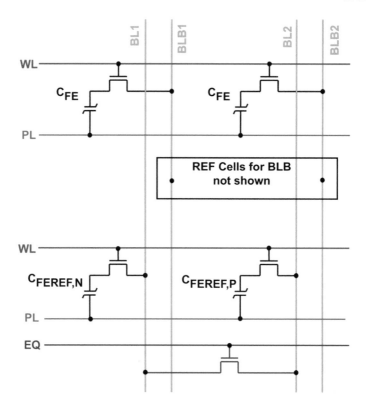

Fig. 3.20 A reference generator with shared bit-line between two neighboring columns [42]

getting the respective BL voltages (V_{BLN} and V_{BLP}), very close to that in the memory cell. Then the two bit-lines are connected by asserting signal EQ (Fig. 3.20) which leads to charge sharing and both the BLs get 0.5 * ($V_{BLN} + V_{BLP}$).

- **Current-Based Referencing**

The techniques that we have discussed so far employ one reference module per column, i.e., one for every sense amplifier. This leads to asymmetry in access of reference and memory cells since access of any row would entail access of the corresponding reference cell. As a result of this, the fatigue in reference FE can be significantly different from the memory FEs. To make the memory-reference accesses more symmetric, array architecture with one reference module per row can be employed [43], as illustrated in Fig. 3.21. Thus, the reference cell is accessed only when the corresponding row is activated. However, since the output of the reference module needs to be connected to multiple sense amplifiers, there is a huge mismatch in the BL capacitance for the reference module and the memory cells. To counter this problem, the BL voltage of the reference module is converted to a current via a properly biased transistor [25], and then, the current mirrors are employed to copy the

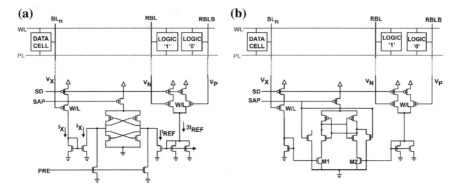

Fig. 3.21 Array architecture with one reference module per row. **a** Conventional symmetrical current-based sensing scheme [34] and **b** modified symmetrical current-based sensing scheme [44]

current to multiple sense amplifiers, as shown in Fig. 3.21. For proper functioning of the sense amplifier, the BL voltage of the memory cell is also translated to current in a manner similar to the reference module. In [44], this technique is further improved by utilizing an n-type feedback transistor in the sense amplifier (Fig. 3.22), which is controlled by the memory BL. If the BL voltage is high, the input current of the sense amplifier is low (since BL voltage controls the gate of a p-type transistor). At the same time, the current in the feedback n-transistor is high, which increases the pull-down strength of the right-hand side of the sense amplifier (connected to the reference current). On the other hand, if the BL voltage is low, the input current is high and the pull-down strength of the reference branch is low. This dynamic feedback increases the sense margin. In [44], it has been shown that this technique leads to ~50% increase in the sense margin.

- **Other Voltage Reference-Based Schemes**

Several other designs for voltage reference exist in the literature, which improves upon some or the other design aspects of the previous techniques. Without going into the details, we will briefly discuss the relevant concepts and point the readers to the associated references. In [45], technique similar to Fig. 3.20 is used with reference module shared between two neighboring columns. However, a node of the reference FE capacitance of one column is connected to the gate of the access transistor of the reference cell belonging to the other column through the word-line. This enables symmetric division of capacitances between the two neighboring reference cells, as detailed in [45], thereby improving the accuracy of reference generation. This work also resets the voltage across the reference FE by controlling the reset transistor via the BL. Since the BL voltage can be high or low, the reference FE capacitances can be made to store positive or negative polarizations randomly, instead of fixed polarization as discussed previously. The design ensures that one of the two reference FE capacitance always has opposite polarization compared to the other. Random

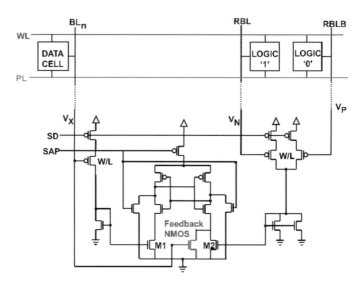

Fig. 3.22 Asymmetrical current-based sensing scheme using a feedback NMOS [44]

polarizations reduce the imprint effects in the reference ferroelectrics. This work achieves sensing time as low as several nanoseconds.

In [46], a bit-line ground sensing scheme is proposed, which aims to mitigate the issue of bit-line capacitance-dependent sense margin. In the schemes discussed previously, the rise in BL voltage (which is a function of C_{BL}) reduces the voltage swing across the FE, thereby reducing the polarization change and in turn, degrading the sense margin [46]. The technique in [46] employs a negative charge tank to clamp the bit-line at ground by injecting negative charge on the bit-line during the read operation. Injecting BL charge changes the output node of the charge tank, which is different for logic states '1' and '0', which can then be used by the sense amplifier to read the data. Thus, the voltage swing across FE becomes less dependent on C_{BL}. The results in [44] show a much reduced sensitivity of sense margin with respect to the bit-line capacitance, which results in up to ~5× improvement in sense margin for high C_{BL}.

- **Sensing Based on Self-referenced Inverter**

Another technique to improve the sense margin and to get around the issue of reference generation is to employ self-referenced inverters [47]. As shown in Fig. 3.23, the BL voltage is fed into an inverter designed with low threshold voltage transistors such that its trip point is between V_{BLP} and V_{BLN}. The trip point is also tuned by employing an independent supply voltage for the inverter. As a result of the trip point being larger than V_{BLP} but smaller than V_{BLN}, the output of the inverter is different for the two logic states. The input and the output of the inverter are compared by the sense amplifier, which amplifies the difference to full swing. Thus, no separate reference is required in this design. The results in [47] show 2.9× to 3.8× improvement

Fig. 3.23 Improving sense
margin utilizing
self-referenced inverters [47]

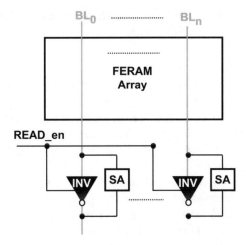

in the sense margin compared to the standard technique of using a reference cell. However, the robustness of this design relies on the inverter trip point being in the middle of the BL voltages corresponding to the two logic states. Since the inverter cannot track the process or temporal variations of the FE capacitances effectively, meeting the design targets for a large array may be challenging.

- **Time Constant-Based Sensing**

While the most common techniques employ BL voltage swing to sense the data in the FERAMs (as discussed so far), another possibility explored in [48] is to detect the capacitance difference by monitoring the time constant associated with BL charging. A current source is employed to charge the BL. Depending on the capacitance (C_{FEP} or C_{FEN}), the time to charge the BL can be significantly different. When the BL voltage exceeds a threshold, the sense amplifier output increases from 0 to V_{DD}. The time at which the sense amplifier step output is generated depends on the logic state. This difference in time is sense by a delay sensing circuit and the output is appropriately latched. In [48], an adaptable circuit is used to generate a reference step signal against which the sense amplifier output is compared by the delay sensing circuit. This adaptable block dynamically changes the time of generation of the reference step signal through feedback from its output such that the rising instant of the reference step signal is mid-way between the rising instants corresponding to two logic states.

As we can clearly see, ferroelectric capacitor-based memories have common read/write paths. Hence, their sensing methods are closely intertwined with consideration of write performance. On the contrary, FEFET-based memories have separate read/write paths (as discussed before). That opens up options to adopt completely different sensing techniques for such memories. We discuss that in the following subsection.

3.3.2 Sensing Techniques for Ferroelectric Transistor-Based Memories

Recalling our discussion on Sect. 3.2, three major versions of FEFET-based memory exist: 2T, 3T, and 4T. In a 2T design (Fig. 3.10a), an FEFET is connected between read word-line and read bit-line in a cross-point fashion. On asserting the read word-line (RWL), low or high current through the FEFET determines its resistance and, in turn, the logic state. To sense, the current read bit-line (RBL) needs to be biased at the virtual ground so that no current flows through the unaccessed cells in the same column.

A separate read access transistor is used in a 3T design (Fig. 3.10b), which isolates the accessed cell from the unaccessed cells. With the assertion of the read word-line (RWL) and application of read voltage on the read bit-line (RBL), current flows through the FEFET, which can be sensed by a conventional current sense amplifier [49] to detect the polarization of the cell. However, both 2T and 3T designs require negative write voltages on the write bit-line (WBL) for switching the FE polarization to a negative value [31].

The 4T design (Fig. 3.10c) avoids negative write voltages (discussed in Sect. 3.2). For the read operation, the two read access transistors are turned ON and depending on the state of the FEFET, the read path exhibits a low or a high resistance. It may be observed that the read path in this design has three access transistors, which reduces the read current. However, the ratio of current between the low-resistance and high-resistance states of the FEFET is enormous ($\sim 10^4$—Fig. 3.5), as a result of which this increase in read path resistance is not that much of a concern. This humungous distinguishability alleviates the design challenges associated with reference generation that we discussed for FERAMs. It is also important to note that in the 4T design, both current sensing and voltage sensing are possible. For the former case, one can apply a fixed read voltage across the read path and sense the current. For the latter case, the read bit-line can be pre-charged. Depending on the state of the cell, either it will remain at high voltage or it will discharge. This difference can be sensed by conventional voltage sense amplifiers [50].

The read operation in FEFET memories is non-destructive, unlike the FERAMs. This is because, the polarization does not need to be switched to detect the states of the memory, as the polarization directly impacts the resistance of the FEFET, which is very easy to sense. While FEFETs look promising to significantly improve the read sensing robustness and to mitigate the read–write conflicts (due to the separation of read–write paths), it must be mentioned that their retention is not as large as FE capacitors. This is due to depolarization fields from the transistor, which reduces the hysteresis of the FEFET compared to the FE capacitor, as discussed in Sect. 3.2. To get a similar hysteresis in the FEFET, their area and/or their thickness needs to be increased. As discussed in [31], this leads to lower write performance at *iso*-energy in FEFET-based memories compared to FERAMs. However, the read benefits of FEFETs are indubitable. It is noteworthy that because of poor retention of FEFETs, they are more suitable for applications that require short-term non-volatility. For such systems, the simplicity of sensing makes FEFET-based memories very attractive.

Having discussed the sensing topologies for ferroelectric memories, we now direct our attention toward the ferroelectric logic circuits. The next section will describe the sensing methods for ferroelectric logic. We will frequently refer to the circuit topologies described in Sect. 3.2 while we describe the sensing methods for a particular topology.

3.4 Sensing Techniques for Non-volatile Ferroelectric Logic

As mentioned in Sect. 3.2, the main focus in designing ferroelectric logic has been directed toward achieving non-volatile flip-flop operation. We have already discussed the backup operation for several non-volatile flip-flop topologies in sect. 3.2, in proximity to circuit diagrams. We refrained from discussion the restore operation as that is closely coupled with the corresponding sensing technique. Here, we complete our discussion by describing the restore operation as well as the sensing mechanism for ferroelectric non-volatile flip-flops.

3.4.1 Sensing Mechanism for Ferroelectric Capacitor-Based Logic

Let us recall the non-volatile flip-flop with a separate backup/restore module shown in Fig. 3.11. With the two FE capacitors exhibiting complementary polarizations, the sensing is performed similar to the 2T-2C FERAM. Thus, during the restore operation, PL voltage is increased, which induces voltage change on the drains of the cross-coupled n-transistors. The FE capacitor storing negative polarization (C_{FEN}) induces a larger voltage on the corresponding branch compared to the other (C_{FEP} with positive FE polarization). Recall, the initial negative polarization switches to a positive value on the application of the PL voltage, while the initial positive polarization remains positive, thus leading to $C_{FEN} > C_{FEP}$. The cross-coupled n-transistors sense this difference and drive the storage nodes of the slave latch to appropriate voltages, thus restoring the state of the flip-flop.

Let us now discuss the restore operation for the non-volatile flip-flop with four FE capacitors in the backup/restore module (Fig. 3.12). Once the power re-appears, the voltage at PL1 is increased while PL2 is kept at 0. The increase in voltage induces a voltage change at nodes X1 and X2 (V_{X1} and V_{X2}), similar to the read operation of the 1T-1C FERAM. The voltage induced is given by

$$V_{X1} = C_{FE1}/(C_{FE1} + C_{FE2})V_{DD} \tag{2a}$$

$$V_{X2} = C_{FE3}/(C_{FE3} + C_{FE4})V_{DD} \tag{2b}$$

Here, C_{FEx} is the capacitance of FEx. Thus, if Q was '1' prior to the power failure, $C_{FE1}=C_{FE4}=C_{FEP}$, which is less than $C_{FE2}=C_{FE3}=C_{FEN}$ (see Fig. 3.2). Thus, V_{X1} is lower than V_{X2}, which is sensed by the sense amplifier to produce a digital output of '1'. This output is used to set the flip-flop state. Similarly, if Q was '0' prior to the power failure, $C_{FE1}=C_{FE4}=C_{FEN}$, which is greater than $C_{FE2}=C_{FE3}=C_{FEP}$. In this case, V_{X1} is greater than V_{X2}, which yields a sense amplifier output of '0'. This output resets the flip-flop. Thus, the four FE capacitances work in a concerted fashion to improve the distinguishability between the two logic states [33]. Compared to the FERAM and the previous non-volatile flip-flop design, where the capacitive division is between FE capacitance and a fixed capacitance (BL capacitance for FERAM and drain capacitance of the cross-coupled n-transistors for the non-volatile flip-flop in Fig. 3.11), this design utilizes the capacitive division between complimentary FE capacitances. This enhances the voltage swing at X1 (X2) when C_{FE1} (or C_{FE3}) $= C_{FEN}$, and suppresses the increase in the voltage at X1 (or X2) when C_{FE1} (or C_{FE3}) $= C_{FEP}$. Moreover, since the complementary flip-flop outputs to control the two pairs of capacitances (FE1,2 and FE3,4) as described above, the voltage swing at the input of the sense amplifier is doubled (similar to the 2T-2C FERAM). Both these factors become instrumental in significantly improving the sensing robustness, thus enhancing the restore operation.

Having discussed the sensing mechanism for FE capacitor-based logic, let us also explore the sensing technique for FEFET-based logic, as discussed next.

3.4.2 Sensing Mechanism for Ferroelectric Transistor-Based Logic

Figure 3.14 showed the flip-flop with a separate backup restore module employing one FEFET in conjunction with other transistors [23]. It may be noted that similar to the FEFET-based memories, the sensing operation is very simple and efficient as the difference in current corresponding to two logic states is quite high. However, it must be ensured that during the restore phase, the backed-up polarization of the FEFET is not disturbed. This is achieved by keeping the V_{GS} of FEFET at 0 by employing other transistors shown in Fig. 3.14. This flip-flop enables restore operation with an order of magnitude lower restore energy than the FE capacitor-based non-volatile flip-flop of Fig. 3.12, at comparable restore delay. Note, in this comparison [23], equal hysteresis of the FEFET and FE capacitor is achieved by employing a larger FE thickness and area in FEFETs (as discussed in Sect. 3.2). In addition, this design offers several other benefits for backup energy, area, leakage etc., which are discussed in detail in [23].

Figure 3.15 showed a topology that integrated FEFETs inside the flip-flop architecture [35]. The restore operation or sensing for this circuit employs the differential FEFETs to drive the storage nodes of the slave latch. As the supply voltage is increased, the FEFET in the low resistance state, pulls-down the corresponding storage node to 0, while other node is pulled up due to the feedback action of the slave latch. Thus, this implementation uses FEFETs in conjunction with the cross-coupled inverters of the flip-flop to restore the state. Similar to the previous non-volatile flip-flop design, the gate-to-source voltage of the FEFETs during the restore operation must be at 0 to ensure that the polarization is not switched as the state of the flip-flop is getting restored.

Compared to the FE capacitor-based non-volatile flip-flops, FEFETs offer clear benefits for the restore operation due to ease of sensing and orders of magnitude difference in the currents corresponding to the two logic states. However, because of lower retention, the applications are limited to the systems that require short-term non-volatility, as we discussed in Sect. 3.2. For such applications, FEFETs offer very interesting design opportunities for non-volatile flip-flops. Further research may uncover new design techniques harnessing the unique attributes of FEFETs for more efficient implementations of non-volatile processors.

3.5 Summary

Ferroelectric (FE)-based devices are very appealing for the design of non-volatile memories and logic. While FE capacitors have been explored for a long time for FERAM designs, the challenges during the read operation have given rise to new schemes for robust sensing. In this chapter, we discussed the pros and cons of several techniques for sensing (step sensing and pulse sensing) and methods to generate accurate voltage references. While most techniques utilize the different in BL voltage swing (which depends on the polarization/capacitance of the FE capacitor), some works have also proposed utilizing the difference in time constants to read the logic state. Alternate designs, which either make sensing self-referenced or with the capability to dynamically tune the voltage reference look very promising to counter the issues associated with FERAM sensing. On the other hand, FE-based transistors (FEFETs) offer appealing features to simplify the sensing of the polarization state, by translating FE polarization to the high or low resistance state. As a result, FEFET-based sensing is non-destructive, unlike FERAMs. While in FEFET-based memories, the robustness of sensing improves significantly compared to FERAMs, it comes at the cost of area. In addition to memory applications, FE capacitors and FEFETs enable the design of non-volatile flip-flops. Although FE capacitor-based non-volatile flip-flops share similar challenges for sensing (or restore operation) as FERAMs, the design issues are not as severe due to relaxed constraints on area and other design factors. Moreover, the feedback action in the flip-flop can be effectively utilized to mitigate the robustness problems associated with detection of FE polarization. On the other hand, FEFET based non-volatile flip-flops are extremely efficient for backup and restore operations and their applications in such circuits look very promising. The following Tables 3.1 and 3.2 highlight qualitative comparison for sensing principles in ferroelectric memory and logic and thereby provide a quick overview of our discussion throughout this chapter.

Table 3.1 Comparison between sensing principles in ferroelectric memory

	Ferroelectric capacitor-based memory	Ferroelectric transistor based memory
1	Common read/write path. Hence, sensing disrupts stored data. (Destructive Sensing)	Separate read/write path hence non-destructive sensing
2	Requires write-back operation hence power hungry	Does not require write-back
3	Relies on the capacitive current difference arising due to difference in average capacitance of ferroelectric capacitor	Relies on the drain current difference arising due to difference in resistance of FEFET
4	Only utilizes the non-volatile feature of ferroelectric material	Combines non-volatility of ferroelectric with gain in the transistor
5	Achievable sense Margin is comparatively lower than FEFET- based memory	Achievable sense margin is higher than FERAMs
6	Satisfactory retention time. Depends on the amount of hysteresis present in ferroelectric polarization-voltage characteristics	Degraded (compared to FERAM) retention time due to depolarization effect of the transistor
7	Leads to better write performance at *iso*-energy compared to FEFET-based memories	Leads to lower write performance at *iso*-energy compared to FERAMs

Table 3.2 Comparison between sensing principles in ferroelectric logic

	Ferroelectric capacitor-based flip-flop	Ferroelectric transistor-based flip-flop
1	The backed-up state stored in the FE capacitors is destroyed during the restore operation (similar to the destructive read of FERAM)	The backed-up state stored in the *gate* of FEFET is not disturbed during the restore operation
2	No write-back is needed in flip-flops (unlike FERAMs). The polarization needs to be programmed only during the next backup operation	No write-back is needed as data do not get disturbed during restore operation
3	FE capacitors require special considerations in terms of layout and fabrication to be integrated with the flip-flop	FEFETs are inherently compatible to be integrated with the flip-flop (based on CMOS transistors). Hence, they incur less fabrication and layout related challenge
4	Restore operation is less efficient (compared to FEFET- based designs) due to moderate distinguishability between logic states	Restore operation is very efficient due to enormous distinguishability between logic states
5	Satisfactory retention time. Depends on the amount of hysteresis present in ferroelectric polarization-voltage characteristics	Degraded (compared to FE capacitor-based flip-flops) retention time due to depolarization effect of the transistor

Acknowledgements The authors thank the DARPA Young Faculty Award program and SRC-GRC for their support for the work on FEFET-based designs.

References

1. Park SP, Gupta S, Mojumder N, Raghunathan A, Roy K (2012) Future cache design using STT MRAMs for improved energy efficiency: devices, circuits and architecture", In: Proceedings of the 49th annual design automation conference (DAC '12). ACM, New York, NY, USA, pp 492–497. http://dx.doi.org/10.1145/2228360.2228447
2. Liu Y, Li Z, Li H, Wang Y, Li X, Ma K, Li S, Chang M-F, John S, Xie Y, Shu J, Yang H (2015) Ambient energy harvesting nonvolatile processors: from circuit to system. In: Proceedings of the 52nd annual design automation conference (DAC '15). ACM, New York, NY, USA, Article 150, 6 p. http://dx.doi.org/10.1145/2744769.2747910
3. Henzler Stephan, Georgakos Georg, Eireiner Matthias, Nirschl Thomas, Pacha Christian, Berthold Joerg, Schmitt-Landsiedel Doris (2006) Dynamic state-retention flip-flop for fine-grained power gating with small design and power overhead. IEEE J Solid-State Circ 41(7):1654–1661. https://doi.org/10.1109/JSSC.2006.873218
4. Ma K, Zheng Y, Li S, Swaminathan K, Li X, Liu Y, Sampson J, Xie Y, Narayanan V (2015) Architecture exploration for ambient energy harvesting nonvolatile processors. In: 2015 IEEE 21st international symposium on high performance computer architecture (HPCA), Burlingame, CA, 2015, pp 526–537. https://doi.org/10.1109/hpca.2015.7056060
5. Fackenthal R, Kitagawa M, Otsuka W, Prall K, Mills D, Tsutsui K, Javanifard J, Tedrow K, Tsushima T, Shibahara Y, Hush G (2014) A 16 Gb ReRAM with 200 MB/s write and 1 GB/s read in 27 nm technology. In: 2014 IEEE international solid-state circuits conference digest of technical papers (ISSCC), San Francisco, CA, 2014, pp 338–339. https://doi.org/10.1109/issc.2014.6757460
6. Dong X, Muralimanohar N, Jouppi N, Kaufmann R, Xie Y (2009) Leveraging 3D PCRAM technologies to reduce checkpoint overhead for future exascale systems. In: Proceedings of the conference on high performance computing networking, storage and analysis, Portland, OR, 2009, pp 1–12. https://doi.org/10.1145/1654059.1654117
7. Lin CJ, Kang SH, Wang YJ, Lee K, Zhu X, Chen WC, Li X, Hsu WN, Kao YC, Liu MT, Chen WC, Lin YC, Nowak M, Yu N, Tran L (2009) 45 nm low power CMOS logic compatible embedded STT MRAM utilizing a reverse-connection 1T/1MTJ cell. In: 2009 IEEE international electron devices meeting (IEDM), Baltimore, MD, 2009, pp 1–4
8. Zwerg M, Baumann A, Kuhn R, Arnold M, Nerlich R, Herzog M, Ledwa R, Sichert C, Rzehak V, Thanigai P, Eversmann BO (2011) An 82μA/MHz microcontroller with embedded FeRAM for energy-harvesting applications. In: 2011 IEEE international solid-state circuits conference, San Francisco, CA, 2011, pp 334–336
9. Endoh T, Koike H, Ikeda S, Hanyu T, Ohno H (2016) An overview of nonvolatile emerging memories—spintronics for working memories. IEEE J Emerg Select Top Circuits Syst 6(2):109–119
10. Chun KC, Zhao H, Harms JD, Kim TH, Wang JP, Kim CH (2013) A scaling roadmap and performance evaluation of in-plane and perpendicular MTJ based STT-MRAMs for high-density cache memory. IEEE J Solid-State Circuits 48(2):598–610. https://doi.org/10.1109/JSSC.2012.2224256
11. Govoreanu B, Kar GS, Chen YY, Paraschiv V, Kubicek S, Fantini A, Radu IP, Goux L, Clima S, Degraeve R, Jossart N, Richard O, Vandeweyer T, Seo K, Hendrickx P, Pourtois G, Bender H, Altimime L, Wouters DJ, Kittl JA, Jurczak M (2011) 10×10 nm^2 Hf/HfOx crossbar resistive RAM with excellent performance, reliability and low-energy operation. In: 2011 international electron devices meeting, Washington, DC, 2011, pp 31.6.1–31.6.4. https://doi.org/10.1109/iedm.2011.6131652

12. Boniardi M, Redaelli A, Cupeta C, Pellizzer F, Crespi L, D'Arrigo G, Lacaita AL, Servalli G (2014) Optimization metrics for phase change memory (PCM) cell architectures. In: 2014 IEEE international electron devices meeting, San Francisco, CA, 2014, pp 29.1.1–29.1.4.https://doi.org/10.1109/iedm.2014.7047131

13. Kohlstedt H, Mustafa Y, Gerber A, Petraru A, Fitsilis M, Meyer R, Böttger U, Waser R (2005) Current status and challenges of ferroelectric memory devices. Microelectron Eng 80, 1 (June 2005), 296–304. http://dx.doi.org/10.1016/j.mee.2005.04.084

14. Mojumder NN, Gupta SK, Choday SH, Nikonov DE, Roy K (2011) A three-terminal dual-pillar STT-MRAM for high-performance robust memory applications. IEEE Trans Electron Devices 58(5):1508–1516. https://doi.org/10.1109/TED.2011.2116024

15. Kawahara Akifumi, Azuma Ryotaro, Ikeda Yuuichirou, Kawai Ken, Katoh Yoshikazu, Hayakawa Yukio, Tsuji Kiyotaka, Yoneda Shinichi, Himeno Atsushi, Shimakawa Kazuhiko, Takagi Takeshi, Mikawa Takumi, Aono Kunitoshi (2013) An 8 Mb multi-layered cross-point ReRAM macro with 443 MB/s write throughput. IEEE J Solid-State Circuits 48(1):178–185. https://doi.org/10.1109/JSSC.2012.2215121

16. Xie Y (2011) Modeling, architecture, and applications for emerging memory technologies. In: IEEE design & test of computers, vol 28, no 1, pp 44–51, Jan–Feb 2011. https://doi.org/10.1109/mdt.2011.20

17. Hoya K, Takashima D, Shiratake S, Ogiwara R, Miyakawa T, Shiga H, Doumae SM, Ohtsuki S, Kumura Y, Shuto S, Ozaki T, Yamakawa K, Kunishima I, Nitayama, Fujii S (2010) A 64-Mb chain FeRAM with quad BL architecture and 200 MB/s burst mode. In: IEEE transactions on very large scale integration (VLSI) systems, vol 18, no 12, pp 1745–1752, Dec 2010. https://doi.org/10.1109/tvlsi.2009.2034380

18. Aziz A, Shukla N, Datta S, Gupta SK (2015) COAST: correlated material assisted STT MRAMs for optimized read operation. In: 2015 IEEE/ACM international symposium on low power electronics and design (ISLPED), Rome, 2015, pp 1–6. https://doi.org/10.1109/islped.2015.7273481

19. Lee YH, Kim HJ, Moon T, Do Kim K, Hyun SD, Park HW, Lee YB, Park MH, Hwang CS (2017) Preparation and characterization of ferroelectric Hf0·5Zr0·5O₂ thin films grown by reactive sputtering. Nanotechnology 28:305703 (13 pp)

20. Lee1 MH, Chen P-G, Liu C, Chu K-Y, Cheng C-C, Xie M-J, Liu S-N, Lee J-W, Huang S-J, Liao M-H, Tang M, Li K-S, Chen M-C (2015) Prospects for ferroelectric HfZrOx FETs with experimentally CET=0.98 nm, SSfor=42 mV/dec, SSrev=28 mV/dec, switch-off <0.2 V, and hysteresis-free strategies. In: 2015 IEEE international electron devices meeting (IEDM), Washington, DC, 2015, pp 22.5.1–22.5.4. https://doi.org/10.1109/iedm.2015.7409759

21. Lee M-H, Chen P-G, Liu C, Chu K-Y, Cheng C-C, Xie M-J, Liu S-N, Lee J-W, Huang S-J, Liao M-H, Tang M, Li K-S, Chen M-C (2015) Prospects for ferroelectric HfZrOx FETs with experimentally CET=0.98 nm, SSfor=42 mV/dec, SSrev=28 mV/dec, switch-OFF <0.2 V, and hysteresis-free strategies. In: IEEE international electron devices meeting (IEDM), Dec 2015

22. Chanthbouala André, Crassous Arnaud, Garcia Vincent, Bouzehouane Karim, Fusil Stéphane, Moya Xavier, Allibe Julie, Dlubak Bruno, Grollier Julie, Xavier Stéphane, Deranlot Cyrile, Moshar Amir, Proksch Roger, Mathur Neil D, Bibes Manuel, Barthélémy Agnès (2012) Solid-state memories based on ferroelectric tunnel junctions. Nat Nanotechnol 7:101–104. https://doi.org/10.1038/nnano.2011.213

23. Wang D, George S, Aziz A, Datta S, Narayanan V, Gupta SK (2016) Ferroelectric transistor based non-volatile flip-flop. In: Proceedings of the 2016 international symposium on low power electronics and design (ISLPED '16). ACM, New York, NY, USA, 10–15, https://doi.org/10.1145/2934583.2934603

24. Shiga H et al (2010) A 1.6 GB/s DDR2 128 Mb chain FeRAM with scalable octal bitline and sensing schemes. IEEE J Solid-State Circuits 45(1):142–152. https://doi.org/10.1109/JSSC.2009.2034414

25. Sheikholeslami A, Gulak PG (2000) A survey of circuit innovations in ferroelectric random-access memories. Proc IEEE 88(5):667–689. https://doi.org/10.1109/5.849164

26. Salahuddin S, Datta S (2008) Use of negative capacitance to provide voltage amplification for low power nanoscale devices. Nano Lett 8(2):405–410
27. Eslami Y, Sheikholeslami A, Masui S, Endo T, Kawashima S (2002) A differential-capacitance read scheme for FeRAMs. In: 2002 symposium on VLSI circuits. Digest of technical papers (Cat. No. 02CH37302), Honolulu, HI, USA, 2002, pp 298–301. https://doi.org/10.1109/vlsic. 2002.1015109
28. Eslami Y, Sheikholeslami A, Masui S, Endo T, Kawashima S (2004) Circuit implementations of the differential capacitance read scheme (DCRS) for ferroelectric random-access memories (FeRAM). IEEE J Solid-State Circuits 39(11):2024–2031. https://doi.org/10.1109/JSSC.200 4.835813
29. Takashima D, Kunishima I (1998) High-density chain ferroelectric random access memory (chain FRAM). IEEE J Solid-State Circuits 33:787–792
30. George S, Ma K, Aziz A, Li X, Khan A, Salahuddin S, Chang M-F, Datta S, Sampson J, Gupta S, Narayanan V (2016) Nonvolatile memory design based on ferroelectric FETs. In: Proceedings of the 53rd annual design automation conference (DAC '16). ACM, New York, NY, USA, Article 118, 6 p, 2016. https://doi.org/10.1145/2897937.2898050
31. Gupta SK et al (2017) Harnessing ferroelectrics for non-volatile memories and logic. In: 2017 18th international symposium on quality electronic design (ISQED), Santa Clara, CA, 2017, pp 29–34. https://doi.org/10.1109/isqed.2017.7918288
32. Wang J, Liu Y, Yang H, Wang H (2010) A compare-and-write ferroelectric nonvolatile flip-flop for energy-harvesting applications. In: The 2010 international conference on green circuits and systems, Shanghai, 2010, pp 646–650. https://doi.org/10.1109/icgcs.2010.5542984
33. Kimura H et al (2013) Highly reliable non-volatile logic circuit technology and its application. Int Symp Mult-Valued Logic 212–218
34. Qazi M, Amerasekera A, Chandrakasan AP (2014) A 3.4-pJ FeRAM-Enabled D Flip-Flop in 0.13-μm CMOS for nonvolatile processing in digital systems. IEEE J Solid-State Circuits 49(1):202–211. https://doi.org/10.1109/JSSC.2013.2282112
35. Li X et al (2017) Advancing nonvolatile computing with nonvolatile NCFET latches and flip-flops. IEEE Trans Circuits Syst I Regul Pap 64(11):2907–2919. https://doi.org/10.1109/TCSI.2017.2702741
36. Benedetto JM, Roush ML, Lloyd IK, Ramesh R (1994) Imprint of ferroelectric PLZT thin-film capacitors with lanthanum strontium cobalt oxide electrodes. In: Proceedings of the 9th international symposium on application of ferroelectrics, pp 66–69
37. Moazzami R, Abt N, Nissan-Cohen Y, Shepherd WH, Brassington MP, Hu C (1991) Impact of polarization relaxation on ferroelectric memory performance. In: Digest of technical papers. 2005 symposium on VLSI circuits, May 1991, pp 61–62
38. Sumi T, Moriwaki N, Nakane G, Nakakuma T, Judai Y, Uemoto Y, Nagano Y, Hayashi S, Azuma M, Fujii E, Katsu S, Otsuki T, McMillan L, de Araujo CP, Kano G (1994) A 256 kb nonvolatile ferroelectric memory at 3 V and 100 ns. In: ISSCC Digest of Technical Papers, pp 268–269
39. McAdams HP et al (2004) A 64-Mb embedded FRAM utilizing a 130-nm 5LM Cu/FSG logic process. IEEE J Solid-State Circuits 39(4):667–677. https://doi.org/10.1109/JSSC.2004.8252 41
40. Miyakawa T, Tanaka S, Itoh Y, Takeuchi Y, Ogiwara R, Doumae AM, Takenaka H, Kunishima I, Shuto S, Hidaka O, Ohtsuki S, and Tanaka S (1999) A 0.5 m 3 V 1T1C 1 Mbit FRAM with a variable reference bitline voltage scheme via a fatigue free reference capacitor. In: ISSCC digest of technical papers, pp 104–105
41. Lowrey TA, Kinney WL (1996) Folded bit line ferroelectric memory device. U.S. Patent 5 541 872, 30 July 1996
42. Wilson DR, Meadows HB (1996) Voltage reference for a ferroelectric 1T/1C based memory. U.S. Patent 5 572 459, Nov 5
43. Papaliolios AG (1993) Dynamic adjusting reference voltage for ferroelectric circuits. U.S. Patent 5 218:566, 8 June 1993
44. Ze Jia, Zhongren Zou, Tianling Ren, Hongyi Chen (2010) An asymmetrical sensing scheme for 1T1C FRAM to increase the sense margin. J. Semicond. 31:115001

45. Jia Z, Zhang G, Liu J, Liu Z, Liou JJ (2014) Reference voltage generation scheme enhancing speed and reliability for 1T1C-type FRAM. Electron Lett 50(3):154–156. https://doi.org/10.1049/el.2013.3193

46. Kawashima S, Endo T, Yamamoto A, Nakabayashi K, Nakazawa M, Morita K, Aoki M (2002) Bitline GND sensing technique for low-voltage operation FeRAM. IEEE J Solid-State Circuits 37(5):592–598, May 2002. https://doi.org/10.1109/4.997852

47. Jia Z, Zhang G, Zhang MM, Ren T, Chen H (2010) A novel fatigue-insensitive self-referenced scheme for 1T1C FRAM. In: 2010 IEEE international memory workshop, Seoul, 2010, pp 1–2. https://doi.org/10.1109/imw.2010.5488402

48. Chandler T, Sheikholeslami A, Masui S, Oura M (2003) An adaptive reference generation scheme for 1T1C FeRAMs. In: 2003 symposium on VLSI circuits. Digest of technical papers (IEEE Cat. No.03CH37408), Kyoto, Japan, 2003, pp 173–174. https://doi.org/10.1109/vlsic.2003.1221193

49. Conte A, Giudice GL, Palumbo G, Signorello A (2005) A high-performance very low-voltage current sense amplifier for nonvolatile memories. IEEE J Solid-State Circuits 40(2):507–514. https://doi.org/10.1109/JSSC.2004.840985

50. Schinkel D, Mensink E, Klumperink E, van Tuijl E, Nauta B (2007) A double-tail latch-type voltage sense amplifier with 18 ps Setup+Hold Time. In: 2007 IEEE international solid-state circuits conference. Digest of technical papers, San Francisco, CA, 2007, pp 314–605. https://doi.org/10.1109/isscc.2007.373420

Chapter 4
Sensing of Phase-Change Memory

**Mohammad Nasim Imtiaz Khan, Alexander Jones, Rashmi Jha
and Swaroop Ghosh**

4.1 Introduction

Phase-change memory (PCM) is a promising non-volatile memory (NVM) that stores data by altering the state of the phase-change material, namely chalcogenide glass. PCM is also known as phase-change RAM (PCRAM) or perfect RAM (PRAM), Ovonic Unified Memory, chalcogenide RAM, and CRAM. PCM offers high integration density, long device endurance, fast access speed, and low read/write power. It is already reported that PCM is faster than flash [1], and its speed is much closer to DRAM. Therefore, PCM is suitable for a wide range of applications and is considered to become one of the mainstream candidates for semiconductor memory. In fact, PCM has already penetrated the semiconductor market as discrete chips. Intel has launched PCM-based solid state drive (SSD) known as Optane [2]. There are several memory series designed with different specifications. So far, the highest capacity offered is 750 GB which belongs to Intel Optane SSD DC P4800X series [3].

In this chapter, we describe the details of PCM technology, followed by description of read/write operation, and finally different sensing techniques.

M. N. I. Khan (✉) · S. Ghosh
Pennsylvania State University, State College, PA, USA
e-mail: muk392@psu.edu

A. Jones · R. Jha
University of Cincinnati, Cincinnati, OH, USA

© Springer International Publishing AG, part of Springer Nature 2019
S. Ghosh (ed.), *Sensing of Non-Volatile Memory Demystified*,
https://doi.org/10.1007/978-3-319-97347-0_4

4.2   Advantages of PCM Over Conventional Memories

As technology is scaling down, keeping the leakage power in tolerable limit has become one of the biggest challenges. Several promising NVMs are being investigated by the scientific community to address the issue. PCM is one of them that has already been commercialized by the semiconductor industry (Intel, [2, 3]). Advantages of PCM over conventional volatile (SRAM/DRAM) or non-volatile (NOR/NAND flash) memory are summarized below:

- **Non-volatility**: PCM is a NVM. Therefore, significant static power can be saved if the memory is powered off during in-active mode compared to SRAM (requires standby power) and DRAM (requires standby power and periodic refresh).
- **Fast read performance**: PCM has the advantage of very low sensing time. The read latency of PCM is comparable to single bit per cell NOR flash, while the read bandwidth is similar to that of DRAM.
- **Fast and low cost write performance**: PCM write operation is faster compared to flash memories. It also operates at lower voltage compared to flash, and thereby provides better write performance.
- **Elimination of erase operation**: PCM does not require erase operation unlike NAND flash. Therefore, PCM is bit-alterable (i.e., the stored information can be switched from '0' to '1', or '1' to '0', without needing an erase step in between).
- **Scalability**: PCM offers high scalability compared to flash memory. Scaling flash (NAND/NOR) is difficult as these memories store information as gate charges, and detection becomes difficult (less electron storage). On the other hand, PCM cell can be scaled significantly and it is also possible to store multiple bit per cell of PCM.

The comparison between PCM and conventional memory technologies is shown in Table 4.1.

Table 4.1 Comparison between PCM with conventional memory technologies [4]

Feature	SRAM	eFlash	PCM
Cell size (F^2)	~120 (6T)	~40 (2T)	~40 (2T)
Read time	<300 ps	1 ms	<20 ns
Write time	<300 ps	10 ns	<1 ns
Endurance	Infinite	10^5	10^7
Leakage	nA/bit	pA/bit	pA/bit
Read energy	Low	Low	Low (5pJ/bit)
Write energy	Low	High	Low
Retention	0	1 year	10 years
Voltage	0.9 V	>3 V	2.5 V
Extra masks	0	7–12	3–5
Non-volatility	No	Yes	Yes

4.3 Basics of PCM

In this section, we present the basics of PCM. We also discuss read/write circuitry for PCM and the corresponding read/write operation.

4.3.1 PCM Cell Description

In the past several years, it has been shown that PCM memory cells can be constructed in several different ways. Each PCM memory design has its advantages and disadvantages when compared to one other. PCM cell designs can be broken down into two primary categories: *heater-based* and *self-heating-based* [5] (Fig. 4.1). *Heater-based* PCM memory cells rely on the existence of a locally placed layer of material to act as a heat reservoir and to heat the adjacent layer of phase-change material [5]. This is commonly done with materials such as tungsten (W) or titanium nitride (TiN) [5–7]. *Self-heating* designs remove the heat reservoir from the design and instead, rely on the internally generated heat within the PCM to cause the material to change the state [5]. Both designs typically rely on commonly used phase-change materials such as $Ge_2Sb_2Te_5$ (GST) [6, 8]; however, other variants exist that use other PCMs such as $In_3Sb_1Te_2$ [9, 10].

Within these two categories of PCM cells, there are multiple different structural methods to design the physical cell. Two of the most common types of PCM cell design are the vertical wall and vertical pillar designs [5]. In both designs, the cell

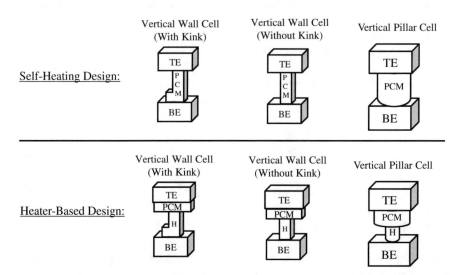

Fig. 4.1 Some common design schemes for *self-heating-* and *heater-based* PCM memory cell designs

is oriented in a vertical fashion with two metal contacts at the top and the bottom of the cell to act as the top and the bottom electrodes for the device. For the vertical wall design, the device layer takes the geometry of a rectangular prism and resides between the two metal contacts. The wall design can also be varied to include a kink in the wall's contact with the bottom electrode to increase its contact area with the bottom electrode's metal [5]. The device layer of the vertical pillar design is exactly like the vertical wall's design; however, the geometry is that of a cylinder instead of a rectangular prism. For *heater-based* PCM memory cells, it is common to have the heating element be smaller than the PCM layer in order to localize the heating process to a particular location within the GST [6]. Other designs for PCM cells such as the planar bridge design also exist, in which a small and very narrow bridge of PCM connects to contacts on either side of the cell [5]. This type of design typically is not as common due to its poor scalability to high-density implementations [5].

4.3.2   Writing to a PCM Cell

Writing to a PCM memory cell is the act of applying a current through the device that causes enough localized heating in order to force the device to either SET or RESET. The SET state within a PCM memory cell is typically a LOW-resistance state where the structure of the phase-change material is crystalline. The SET state is usually achieved by keeping the PCM device above its melting temperature (T_{melt}) for a long enough period of time to allow the device to enter the crystalline state. If the device is instead quickly quenched by quickly removing the current through the device during a shorter heating process, the crystal structure inside the device instead becomes amorphous. This process places the device in its RESET state. This causes the internal resistance of the device to be a much higher value, typically at least one order of magnitude higher or more [5].

To determine the rate at which the crystallization of the PCM device occurs, a commonly accepted temperature-dependent set of Eqs. (4.1) and (4.2) can be used to find this rate [11]. Descriptions for each variable in the Eqs. (4.1) and (4.2) can be found in Table 4.2.

$$v_g(T) = \frac{4 r_{\text{atom}} k_B T}{3\pi \lambda^2 R_{\text{hyd}} \eta(T)} \left(1 - e^{\frac{-\Delta G(T)}{k_B T}} \right) \tag{4.1}$$

$$\Delta G(T) = \Delta H_m \left(\frac{T_{\text{melt}} - T}{T_{\text{melt}}} \right) * \left(\frac{2T}{T_{\text{melt}} + T} \right) \tag{4.2}$$

Arriving at a value for $v_g(T)$ can help determine a PCM's switching speed, as the switching speed is largely governed by this crystallization rate (typically in the order of tens or hundreds of nanoseconds depending on the design of the device [8, 12] and sometimes even longer [11]). However, some newer designs of PCM cells have lowered the switching times for the SET and the RESET states to the range of

Table 4.2 Variable descriptions for Eqs. (4.1) and (4.2)

Variable	Description
$v_g(T)$	Temperature-dependent crystallization rate of the PCM
r_{atom}	Atomic radius of PCM
k_B	Boltzmann constant
λ	Diffusional jump distance in PCM
R_{hyd}	Hydrodynamic radius of PCM
$\eta(T)$	PCM's viscosity
$\Delta G(T)$	Difference between liquid (melted) and crystalline phase's Gibbs energy for material
ΔH_m	Heat of fusion
T_{melt}	Melting temperature of PCM

Table 4.3 Variable descriptions for Eq. (4.3)

Variable	Description
R_{SET}	Set resistance
I_{prog}	Programming current
$\rho_{el}^{GST-OFF}$	Electrical resistivity for PCM material in SET state
ρ_{el}^{GST-ON}	Electrical resistivity for PCM material in RESET state
$\rho_{th}^{GST-OFF}$	Thermal resistivity for PCM material in RESET state
ΔT	Change in temperature (Fourier's Law)

10–50 ns, which brings PCM memory cells a bit closer to the level of the performance of competing memory technologies such as DRAM [12].

The state transition event from SET to RESET is also an important factor. The transition between the two states not only determines the current required to change the state of the device, but also determines other factors such as power consumption or read latency [5]. For both *self-heating-* and *heater-based* types of PCM cells, a relationship product for R_{SET} (set resistance) and I_{prog} (programming current or current through the PCM device) can be derived based upon Fourier's Law, the Joule effect, and both thermal and electrical resistances [5]. The relationship product for *self-heating* PCM cell designs can be described by Eq. (4.3) with all the variables described in Table 4.3 [5].

$$\text{Self-Heating}\left(R_{SET} * I_{prog}\right) = \rho_{el}^{GST-OFF} \sqrt{\frac{2 * \Delta T}{\rho_{th}^{GST-ON} * \rho_{el}^{GST-ON}}} \qquad (4.3)$$

Table 4.4 Variable descriptions for Eq. (4.4)

Variable	Description
R_{SET}	Set resistance
I_{prog}	Programming current
X	Dimension parameter
ΔT	Change in temperature (Fourier's Law)
ρ_{el}^{H}	Electrical resistivity of heater
$\rho_{\text{th}}^{\text{GST}}$	Average thermal resistivity for PCM material
$\rho_{\text{el}}^{\text{GST-OFF}}$	Electrical resistivity for PCM material in RESET state
ρ_{th}	Thermal resistivity
η	Dimension parameter

The resistivity (along with other properties) of the heating element must also be taken into account when finding a proper R_{SET} and I_{prog} relationship for a *heater-based* type of PCM memory cell. This can be seen in Eq. (4.4) [5]. All variable descriptions for the *heater-based* relationship equation can be found in Table 4.4.

$$\text{Heater-Based}\left(R_{\text{SET}} * I_{\text{prog}}\right) = \sqrt{\frac{\rho_{\text{el}}^{H} * \Delta \text{T}}{\rho_{\text{th}}^{\text{GST}} * X * \eta}} + \rho_{\text{el}}^{\text{GST-OFF}} \sqrt{\frac{X * \Delta \text{T}}{\rho_{\text{th}} * \rho_{\text{el}}^{H} * \eta}} \quad (4.4)$$

Minimizing factors such as programming current are often key engineering challenges as it can affect characteristics such as device endurance. PCM devices have previously shown high-level endurance for currents as high as $600\,\mu\text{A}$ for more than 10^{10} SET-RESET cycles [8]. They are also able to hold a single state for a period of 300 years at an ambient temperature as high as $85\,^{\circ}\text{C}$ [8].

4.3.3   Description of Read/Write Circuitry

The basics of physical memory configuration for any standard digital memory system come into play when configuring a block of PCM cells to be used as a piece of read and write accessible memory. A basic method of setting up an individual PCM cell to store a single bit of information is in the 1T1P configuration, where the PCM cell is placed on the drain of a MOSFET with the PCM cell's bottom electrode grounded [12]. A diagram depicting such a configuration can be seen at the bottom of Fig. 4.2. The source of the MOSFET can then be attached to a bitline, and the gate of the MOSFET can be connected to a word enable signal [12]. The 1T1P cells can then be arranged in a grid format to create a 2D array of 1-bit memory cells. The cells can then be connected to create a set of *m* words. Each word is created by connecting a

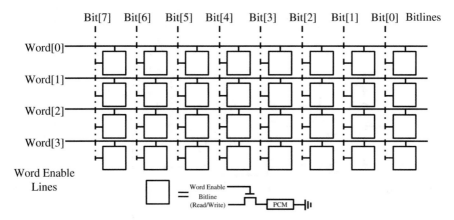

Fig. 4.2 Example diagram of a four-word PCM memory block where each word contains eight bits. Each square within the array represents a single 1T1P memory cell (as shown at the bottom of the figure)

row of n cells together with a single word enable signal. An array of separate word enable signals evolves from this design where giving any word enable line a HIGH logic value allows the user to either read/write to the entire word. The information to be written/read to and from each word is sent along the n bitlines in the array. By connecting the ith bit of each word together into a set of n bitlines, any word can be easily accessed along the bitlines due to the word enable signals. An example circuit demonstrating such a setup can be seen in Fig. 4.2 where a four-word block of PCM memory can be seen with each word containing 8 bits.

Additional circuitry is needed to get the PCM memory block of words to interact with the outside devices in a useful manner. Following the basics of memory access architecture design, the word enable lines created within the PCM memory block can be connected to the output of a $log_2(m) \times m$ decoder, where the input of the decoder is the binary address of the desired word to be accessed. The decoder would also have a read/write enable line (can just be an OR'ed output of the read and write signals or a separate signal) to ensure words are only being accessed when instructed. If the memory block begins to possess a large volume of words to where the architecture for the decoder begins to become cumbersome, other more advanced decoding could be used such as coincident decoding. This decoding technique would reduce the amount of hardware required to handle larger blocks of memory.

The bitlines have a set of controlled buffers applied to them to allow write and read access to the memory block [12]. The input controlled buffers either transport a value desired to be written to a specific word or the memory block's designated read voltage [12]. The output controlled buffers would transport the read output values from the memory block to additional circuitry that would interpret the read voltages into 0's and 1's (see the *sensing schemes* section for more information on this circuitry) [12]. A diagram depicting this setup of circuitry can be seen in Fig. 4.3.

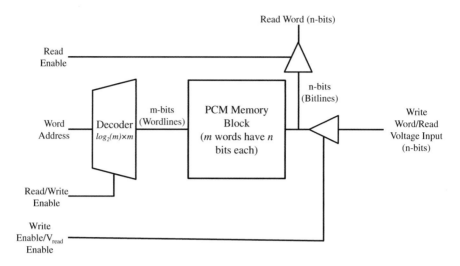

Fig. 4.3 Diagram showing a basic circuit configuration to access words within the PCM memory block

4.3.4 Read/Write Operation

Writing a new value to a word in PCM is similar to typical digital memory, with one critical difference. When writing, the write enable signal to the bitlines along with the read/write enable to the decoder would both go HIGH as expected. The binary address is also sent into the decoder with no special caveats. The difference lies in the signals transmitted along the bitlines to the PCM block. Instead of these values being the usual HIGH- and LOW-voltage values, these will either be SET (1) or RESET (0) pulses to properly configure the crystal structure of each PCM device [5, 12]. As previously mentioned, SET pulses typically last longer than RESET pulses, so a set of pulses of two different lengths will enter the memory block to configure the specified word to the proper binary value.

During the read operation, the goal is to obtain the current state of the device (SET or RESET) without affecting the device's state. To do this, a low V_{read} voltage is passed via the bitlines to each PCM cell within a specific word [12]. The voltage of the bitlines will change in accordance with the resistance state of each PCM device when the word is read. If the PCM device is in the SET state, the voltage for that bitwill be a low value [12]. If the device is instead in the RESET state, the voltage will be higher than it would be in the SET state [12]. External circuitry beyond the access hardware then interprets these read voltage values into 0's or 1's (see the *sensing schemes* section for elaboration on this circuitry).

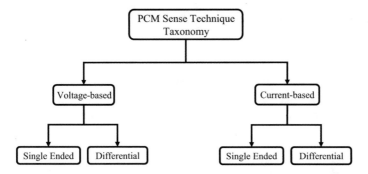

Fig. 4.4 PCM sense technique taxonomy

4.3.5 Challenges for PCM Devices

PCM devices suffer from a few design obstacles, including high write current and long switching times [13]. In addition to these design obstacles, PCM devices face a set of common faults that they can experience during certain use cases. For example, an incorrect value can be read from the PCM device if the device is read too soon after a write operation. This is because a settling time is required after a PCM device has experienced a write operation to allow the internal crystalline/amorphous state of the material to stabilize [14]. Read current has also been shown to affect the state of the PCM cell and cause it to degrade due to an internal buildup of defects within the atomic structure [15]. Furthermore, the resistance of PCM drifts over time [16]. The resistance drift issue is explained in Sect. 4.5.1.

One of the most common faults that occur within PCM device is a 'Stuck at 0' fault [17]. The cause of this fault is most often attributed to overuse of the device, which will cause its layer of PCM material to separate from its metal top electrode layer, making it impossible to further write to the device [17]. Read operation will detect a HIGH-resistance state due to the physical separation of the PCM and the top electrode. Therefore, the device is permanently 'Stuck at 0.' It should be noted that, HIGH-resistance state is considered as data 0 [17] in this paragraph; however, in the rest of the Chapter, HIGH-/LOW-resistance states are considered as data 1/0 respectively.

4.4 Sensing Schemes

The data stored in PCM is read out as the resistance of the bit cell. If a selected cell is in a crystalline (amorphous) state, the resistance of the cell is LOW (HIGH). Sensing the resistance of PCM cell can be divided into various categories as illustrated in the taxonomy in Fig. 4.4.

Broadly, the PCM sensing can be divided into:

(i) Voltage-based sensing;
(ii) Current-based sensing.

Both of the above techniques can be further subdivided into single-ended sensing and differential sensing based on the presence of specific reference voltage. In the following subsection, details regarding each sensing technique are given.

4.4.1 Voltage-Based Sensing

4.4.1.1 Voltage-Based Single-Ended Sensing

Figure 4.5a shows a basic circuit for voltage-based single-ended sensing. First, both V_{ref} and V_{data} are pre-charged to V_{dd}, and then, both reference cell and data cell are accessed by asserting corresponding wordline signals (Fig. 4.5b, c). Both V_{Ref} and V_{data} start to discharge. However, if $R_{Data} > R_{Ref}$ (data $= 1$), V_{Ref} discharges faster compared to V_{data} and tries to pull V_{data} to V_{dd} by turning ON PMOS P_2 (Fig. 4.5b). On the other hand, if $R_{Data} < R_{Ref}$ (data $= 0$), V_{data} discharges faster compared to V_{Ref} and tries to pull V_{Ref} to V_{dd} by turning ON PMOS P_1 (Fig. 4.5c). It should be noted that V_{data} is the output voltage ($V_{data} = V_{Out}$) and represents the state of the memory. Although the sensing circuit includes a reference cell, the voltage V_{Ref} is not a constant voltage which can be considered a reference voltage. Therefore, this technique is called voltage-based single-ended sensing.

4.4.1.2 Voltage-Based Differential Sensing

Figure 4.6 shows a basic circuit for voltage-based differential sensing. A constant current, I_{CELL} is injected into the memory cell by asserting WL signal. V_{CELL} is zero before the sensing operation. However, I_{CELL} charges this node based on the PCM cell resistance. If PCM cell resistance is HIGH (data $= 1$), $V_{CELL} > V_{REF}$ and thereby the sense amplifier (S/A) outputs 1 ($D_{Out} = 1$). However, if PCM cell resistance is LOW (data $= 0$), $V_{CELL} < V_{REF}$ and the sense amplifier (S/A) outputs 0 ($D_{Out} = 0$). Therefore, the resistance states of PCM cell can be easily determined using a high-speed comparator and by setting a reference voltage V_{REF} at the middle of voltage sensing gap (gap between values of V_{CELL} for data 0 and data 1).

Above description is a general voltage-based differential sensing technique suitable to any memory technology. However, many PCM-specific improvements have been proposed in prior research works to improve the sensing performance and

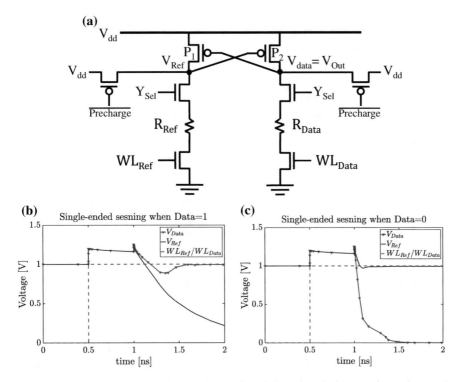

Fig. 4.5 **a** Circuitry for simple voltage-based single-ended sensing; timing waveform of V_{ref} and V_{out} when **b** data $= 1$; and **c** data $= 0$

Fig. 4.6 Voltage-based differential sensing

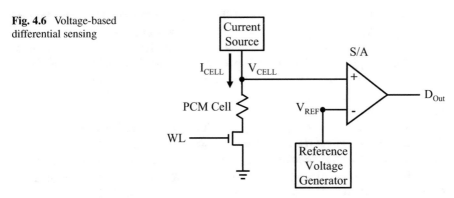

the robustness. The state-of-the-art PCM-specific voltage-based differential sensing techniques are summarized below:

A. *Auxiliary current for faster sensing* [18]:

Voltage-based sensing which requires a long time to charge the bitline due to associated large load. Therefore, this work [18] proposes to implement an auxiliary

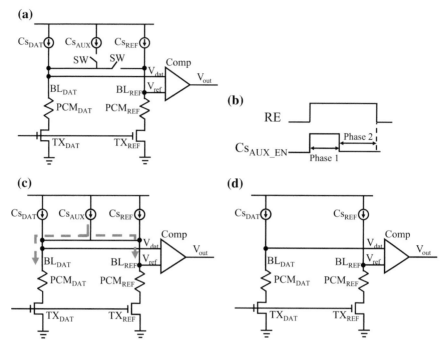

Fig. 4.7 **a** Auxiliary current-based fast sensing circuit [18]; **b** waveform for RE and Cs_{AUX_EN} signals; equivalent circuit during **c** phase 1; and **d** phase 2

current source for a specific window of time to charge the data/reference bitlines faster (Fig. 4.7a). This sensing circuit has two phases of operation which are implemented with two control signals, namely read enable (RE) and auxiliary current enable (Cs_{AUX_EN}) (Fig. 4.7b). RE is kept enabled during the entire duration of read operation, while Cs_{AUX_EN} is enabled to close the SW switches and to enable the auxiliary current source Cs_{AUX} during phase 1. Brief description of the two operational phases is given below:

Phase 1 (accelerated charging stage): In this phase, the switches SW are closed and auxiliary current source is enabled by asserting Cs_{AUX_EN} (Fig. 4.7c). Therefore, the auxiliary current source (Cs_{AUX}) provides auxiliary current (I_{aux}) to the data/reference bitlines (BL_{DAT}/BL_{REF}) of the storage memory device PCR_{DAT}/the reference memory device, PCR_{REF}. The time required to charge the bitlines is reduced due to the auxiliary current (I_{aux}). Optimum duration of phase 1 can be selected based on the loads of the bitlines.

Phase 2 (differentiation stage): In this phase, the switches SW are open and the auxiliary current source Cs_{AUX} is disconnected from the data/reference bitlines (Fig. 4.7d). Then, V_{dat} and V_{ref} gradually stabilize. The comparator detects the storage state of the storage memory device PCR_{DAT} according to the voltage levels V_{dat} and V_{ref}.

It should be noted that a delay element and the read enable signal RE can be utilized to generate the auxiliary current source enable signal Cs_{AUX_En}.

B. *Reducing read latency* via *Early and Turbo Read* [19]:

This work proposes two techniques, namely *Early Read* and *Turbo Read* for reducing read latency. The basic concepts are given below:

Early Read: Early Read makes the sense amplifier latch the data early, thereby reduces the sensing time. This is achieved by reducing the reference resistance. However, this would cause a few cells in the SET state (data = 1) to be classified as RESET (data = 0) which leads to read failure. Therefore, the system needs to be able to tolerate a larger bit error rate (BER) (~several parts per million) which requires a strong Error Correcting Code (ECC) (that can correct almost six or more bits per line). The work incorporates Berger code [20] that efficiently detects all unidirectional errors in a data block at the very little expense of consuming storage overhead and complexity.

Turbo Read: Turbo read increases the read voltage to achieve reduced retention time. However, increasing read voltage increases the probability of read disturb. Therefore, this work proposes to increase the read voltage higher than normal but low enough so that the read disturb can be avoided/corrected in a practical and cost-effective manner. The PCM characterization data in [21] shows that an increase in sensing voltage by 30 mV increases the likelihood of read disturb errors by three orders of magnitude. This work increases the sensing voltage from 0.70 to 0.79 V, which would increase the read disturb probability for a bit from 10^{-18} (baseline read disturb probability) to 10^{-9}. The work also analyzes the area/power and complexity associated with an ECC to mitigate the error correcting code.

Early Read requires memory system to support two types of read latency (low/normal latency). However, Turbo read does not require a dual latency support from the memory system as it targets the sensing voltage as a means to reduce the sensing time.

C. *Multi-level sensing in PCM* [22]:

This work proposes a new multi-bit PCM architecture with differential sensing which provides 14–50% increase in information storage density over single-bit PCM with single-ended sensing and approximately equivalent sense margins.

In the proposed architecture, data is stored in groups of n-bit cells where every one of the n possible states appears exactly as one of the n cells. Differential sensing enables reading collective state of the group, and thereby each group of n-bit cells is read at the same time. For example, if each cell contains three states and a group of three cells is collectively encoded, six collective states are possible. Three comparators can produce three outputs to indicate the collective state of that particular group. It should be noted that each collective state corresponds to a unique set of digital outputs. However, not every combination of digital outputs corresponds to a possible collective state.

Each group provides its own reference as it is predetermined that each group contains all possible states. Therefore, the possibility of reference behaving differently

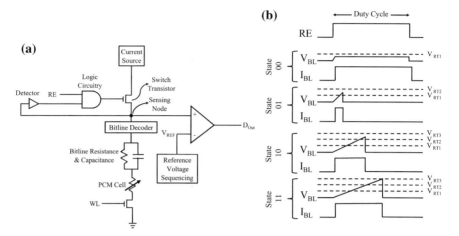

Fig. 4.8 **a** Proposed circuitry for current compliance sensing; **b** timing waveform of RE, bitline voltage V_{BL}, and bitline current I_{BL} [23]

from storage cells is nullified. Fully differential sensing also enables comparing every pair of outputs (in a given group) against each other.

It should also be noted that collective states of three three-state cells contain more information than four 1-bit cells could. Therefore, the proposed architecture ensures high density with the advantage of differential sensing.

D. *Current complaint sensing* [23]:

It has been mentioned before that the read current can change the state of the cell if the current is high enough to generate sufficient thermal energy. In [23], a current compliant sensing technique is proposed which controls the pulse width of the sensing current. Read current from a constant current source is applied to the cell that is being read. A logic circuit is implemented to control the read current pulse duration in response to the PCM cell resistance. Figure 4.8a shows the circuit implementation, and Fig. 4.8b shows the corresponding waveforms.

During read operation, read enable (RE) signal is applied to the logic circuitry (modeled by an AND gate) shown in Fig. 4.8a as well as wordline/bitline/sourceline signals to select a cell. The logic circuitry has another input which is active LOW and coupled to the output of the detector (modeled by a NOT gate). The input of the detector is coupled to the sensing node (Fig. 4.8a) or D_{out} (in another implementation proposed in this work). Sense amplifier compares the voltage of sensing node with V_{ref} and outputs D_{out} which corresponds to a data state of the PCM cell. For a multi-bit cell, the reference voltage V_{ref} is supplied by reference voltage sequencing circuits changed in a sequence as V_{ref1}, V_{ref2}, V_{ref3} (enables of detection of four data states). V_{ref1} is used for distinguishing between the lowest resistance phase and a first intermediate phase, V_{ref2} is used for distinguishing between the first intermediate phase and a second intermediate phase, and V_{ref3} is used for distinguishing between the second intermediate phase and the highest resistance phase (Fig. 4.8b).

The sequencing circuit can be implemented using voltage dividers and a set of switches under the control of a read state machine. The detector operates with a single trigger level corresponding to V_{REF3} which can be used to ensure that the voltage across PCM cell never exceeds the switching threshold voltage. The detector logic outputs a logic 1 when the sensing node reaches a trigger voltage, and the switch transistor is turned off, disconnecting the current source from the sensing node and thereby ending the read cycle. The work also proposes three detectors coupled in parallel with respective trigger levels to be enabled in the same sequence as the reference voltage is applied and getting D_1, D_2, and D_3 instead of a single D_{out}.

E. *Reconfigurable sensing for variable storage level* [24]:

This work presents a novel reconfigurable sensing technique for Multi-level Cell (MLC) PCM with the flexibility to change reading precision of analog resistance levels.

The basic concept of the proposed sensing technique is depicted in Fig. 4.9. It employs a single dynamic reference voltage and fixed WL voltage level unlike prior MLC sensing techniques for NAND flash using either multiple fixed reference voltages (single WL voltage level) [25] or multiple WL voltage levels (single fixed reference voltage) [16]. Therefore, the technique eliminates complicated multiple reference circuits and expensive DAC and provides flexibility for changing sensing precision during read operation. It should be noted that the scheme can also be applied as a built-in self-test (BIST) unit to read the analog information of single-level cell (SLC) PCM with adjustable reference value which dramatically reduces test time/cost. The operating principle is summarized as follows:

(1) Sensing time is recorded by a clock cycle counter;
(2) Reference voltage is generated dynamically. The generated voltage is ramping up continuously with time as sensing proceeds;
(3) Sense amplifier output switches when the reference voltage crosses the sensed cell voltage. The corresponding sensing time is recorded in a counter and that value is latched into output registers. This stored value in the output register is used as the sensed resistance level.

It has been reported that this work provides 8-bit precision (adequate for 7b/cell PCM, i.e., 128 resistance levels) with a read access latency of 5 μs (measured at 50 MHz clock). Therefore, it achieves higher performance compared to 35–50 μs in state-of-the-art 2b/cell NAND flash.

4.4.1.3 Voltage-Based Sensing Using Delta-Sigma Modulation [26]

MLC PCM resistance varies from a few kΩ to several kΩ or even MΩ [27, 28]. Traditional sensing schemes use differential sensing. However, differential sensing

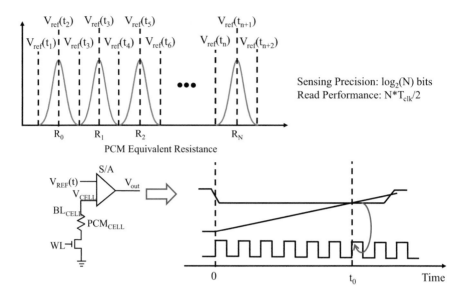

Fig. 4.9 Reconfigurable sensing technique for multi-level cell (MLC) PCM [24]

is sensitive to process variation and noise and requires wide resistance margins to sense reliably. Therefore, this work proposes delta–sigma modulation-based (DSM) four sensing topologies which employ voltage-based sensing. These techniques are robust [29–31] and faster compared to traditional techniques.

Basics of DSM sensing circuit:

An example of DSM sensing circuit is shown in Fig. 4.10. The circuit consists of an integrating bitline capacitor C_{bit}, an analog-to-digital converter (ADC) (comparator connected to the bitline) and a feedback loop which acts as a digital-to-analog converter (DAC). The DAC keeps the bitline voltage constant.

First, the bitline is charged to a read voltage (greater than comparator reference voltage), and a cell is selected to sense its content. The comparator outputs logic 1 (read voltage is greater than reference voltage) which turns off the PMOS M3. However, the bitline discharges through the selected cell based on its content (discharges faster if the cell is in LOW-resistance state). The comparator outputs logic 0 when the bitline discharges below the reference voltage. This turns ON the PMOS M3, and the bitline capacitor charges back to read voltage which (greater than the reference voltage) in turn makes the output of comparator logic 1 and turns OFF the PMOS. The sensing circuit senses the change in bitline voltage over a period, and the number of times (M) the output of the sense amp goes HIGH/LOW in this period is recorded using a counter. This count is further compared to a reference count [32] which is selected to represent a logic 0/1 (or other levels in case of MLC). Therefore, a count value greater than the reference count indicates one logic state, and a count value less than the reference count indicates the another.

Fig. 4.10 DSM sensing
circuit [26]

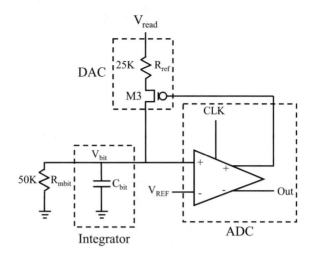

The reference resistor used in the DSM sensing circuit can either be a discrete resistor or a switched-capacitor-based resistor. Furthermore, the comparator can have zero or nonzero offset (for switching point voltage). Based on these observations, this work proposes four topologies:

(1) Reference resistor-based delta–sigma sensing with no offset
(2) Reference resistor-based delta–sigma sensing with offset
(3) Switched-capacitor resistor-based delta–sigma sensing with no offset
(4) Switched-capacitor resistor-based delta–sigma sensing with offset.

The voltage across the memory cell during read can cause the memory cell to switch. Therefore, it is desirable to reduce the voltage across the cell which can be achieved by a comparator with a built-in offset on its reference terminal. This helps in minimizing the number of reference voltages used in the sensing circuit. On the other hand, a simple reference resistor has the disadvantage of restricting the flexibility for tuning its resistance value once they are designed on chip, whereas a switched-capacitor resistor can be tuned by varying the clock supplied to it. Therefore, a switched-capacitor resistor with a built-in offset in the comparator allows the most flexibility during operation.

The advantage of this work is the higher accuracy of the sense circuit due to this averaging effect which removes the impact of unwanted signals like noise. A wide range of resistances can also be sensed using the four sensing topologies proposed in this work.

4.4.2 Current-Based Sensing

4.4.2.1 Current-Based Single-Ended Sensing

The advantage of current-based single-ended sensing over current-based differential sensing is that it does not require generation of robust reference current [33, 34]. Reference current is chosen very carefully considering the cell current fluctuation due to process variation. This process can increase the test time [35]. Furthermore, the speed and performance of differential sensing decrease as the supply voltage is scaled down.

In [35], a current-based single-ended sensing is proposed for low-voltage embedded EEPROM. However, the idea can also be extended to PCM. The basic idea is to apply a constant V_{read} voltage to a cell that is being read and pass current through the bit cell that is sensed. If the current is greater than a trip point, the read circuitry converts it to a logic 0. However, if the current is shorter than the trip point, the read circuitry converts it to a logic 1. The work proposes to tune the trip point by sizing the transistor properly. It should be noted that the proposed work shows a good sensing speed even for a low level of sense current (1 μA). Furthermore, this work performs well for low supply voltage and can be implemented for any kind of NVM.

4.4.2.2 Current-Based Differential Sensing

In [35], an example of current-based differential sensing technique for NVMs is presented (Fig. 4.11). In this technique, a small read voltage V_{READ} is applied to the source of both NMOS N_1 and N_2. The PCM cell which needs to be sensed is selected by asserting appropriate bitline/worline/sourceline resistances. The selected PCM cell and reference cell are clamped at the source nodes of the access transistors, N_1 and N_2, respectively, using a high gain amplifier (Fig. 4.11). Reference cell generates I_{Ref} where selected PCM cell generates I_{Data}. Therefore, I_{Data} can be sensed out from data bitline, and I_{CELL} and I_{REF} are compared using simple current comparator to determine the PCM's resistance state. If the cell is in HIGH state, I_{Data} is smaller than I_{REF} and the current comparator needs to generate 1 as output. However, if the cell is in LOW state, I_{Data} is greater than I_{Ref} and the current comparator needs to generate 0 as output. I_{Ref} should be set in the middle point of the read currents of HIGH-resistance cell (I_{Data_High}) and LOW-resistance cell (I_{Data_Low}) in order to maximize the current sensing margin. It should be noted that V_{Read} should be small enough such that I_{Data} must not exceed the critical current in order to avoid read disturb. However, sense margin reduces if V_{Read} is reduced. Therefore, appropriate V_{Read} needs to be selected considering this trade-off.

Fig. 4.11 Current-based
differential sensing. The
circuit is a simplified version
of the example presented in
[35]

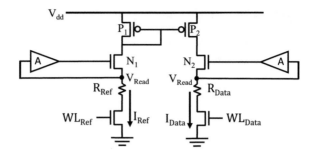

4.4.3 Comparison of Different Sensing Techniques

Detailed study has been performed on voltage-based sensing to address PCM-specific sensing issues. However, current-based sensing scheme is also widely adopted for NVMs. In fact, current-based sensing scheme is usually better for designs with long BL arrays and small PCM cell currents. The reason is current sensing amplifier is more immune to bitline noises. Furthermore, current sensing amplifier is faster compared to voltage-based sensing. However, it incurs larger area [36] compared to voltage-based technique. Therefore, appropriate sense technique should be implemented based on the target specification and the application of the memory.

4.5 Sensing Issues

4.5.1 Resistance Drift

The equivalent resistance of a PCM cell drifts (increase) from its designed value over time. The primary reason of this drifting is caused by rearrangement at atomistic level (defect annihilation) of the disordered structure and the change in the density of localized states accompanied by an energy release [37, 38]. The consequence is the material evolves into a thermodynamically stable state and thereby changing the electrical properties of the chalcogenide material, i.e., increasing threshold voltage and resistance [16]. It should be noted that this process is stochastic in nature as the process is directly linked to the defects in the amorphous structure after programming. Therefore, it is difficult to design sense circuitry considering this issue and avoid read errors.

The work in [16] proposed a technique which takes the drifting into account and generates a time-aware sensing scheme. It delivers a method to calculate the threshold between intermediate resistance levels, taking into account the experimental statistical data of drifting behavior at various initial resistance levels. The technique is modeled considering an assumption that both the resistance at amorphous state and the resistance drift follow Gaussian distributions. The work details the method to calculate an adaptive time-aware resistance threshold that generates the minimum reading errors. This threshold gives the value of the quantized reference resistance between two consecutive levels (with respect to time) according to the Gaussian distribution, and it is further used as a reference for comparison during read.

4.5.2 High Sensing Time

PCM resistance varies from KΩ to MΩ. Therefore, sensing time can be significant (especially sensing HIGH state). Auxiliary current sources can be implemented to charge the bitline voltage faster as proposed in [18]. This reduces the sensing time. Higher read voltage can also be applied [19] to reduce the sense time. However, this increases the probability of read disturb. Another approach is to make the sense amplifier latch the data early [19]. This increases read failure which could be mitigated by implementing ECC.

4.5.3 Sensing Time Variation

PCM sensing time depends on the value of HIGH and LOW resistances. The resistance of both states varies with process variation and temperature. Therefore, a guard band should be incorporated with the target sense time. This makes sure that the weakest bit at the worst operating condition can also be sensed.

4.5.4 Multi-level Sensing

PCM offers MLC functionality. This is achieved by exploiting the fact that multiple resistance level can be obtained in the wide resistance range between the HIGH and LOW states. However, this calls for designing new circuits which could sense different resistance levels. The work in [24] proposes a sensing technique which provides the flexibility to change reading precision of analog circuits. DSM sensing technique proposed in [26] also enables robust sensing for PCM MLC.

4.6 Conclusion

In this chapter, we presented the basics of PCM and described the different design techniques for PCM cell. We also discussed the basics of read and write operation of PCM along with the required read/write circuitry. We explained voltage-based and current-based PCM-sensing methodology both of which can be designed as single-ended or differential sensing. Sensing strategy for a memory technology plays a vital role in determining the performance of the memory. Voltage-/current-based single-ended/different techniques are common for all traditional and emerging memory technology. However, the techniques require modification to address the memory specific issues. Therefore, in this chapter, we have summarized the different state-of-the-art sensing schemes which are specific to PCM. We have also discussed different sensing issues related to PCM and possible countermeasures.

References

1. https://www.theverge.com/2016/5/17/11693054/ibm-phase-change-memory-breakthrough-ram-flash-storage
2. https://ark.intel.com/products/97544/Intel-Optane-Memory-Series-16GB-M_2-80mm-PCIe-3_0-20nm-3D-Xpoint
3. https://www.intel.com/content/www/us/en/products/memory-storage/solid-state-drives/data-center-ssds/optane-dc-p4800x-series/p4800x-750gb-2-5-inch.html
4. https://www.eetimes.com/document.asp?doc_id=1332517
5. Boniardi M et al (2014) Optimization metrics for phase change memory (PCM) cell architectures. Electron Devices Meeting (IEDM)
6. Russo U et al (2008) Modeling of programming and read performance in phase-change memories-Part I: cell optimization and scaling. IEEE Trans Electron Devices 506–514
7. Servalli G (2009) A 45 nm generation phase change memory technology. IEEE International, pp 1–4
8. Pellizer F et al (2004) Novel μtrench phase-change memory cell for embedded and stand-alone non-volatile memory applications. Symposium on VLSI Technology Digest of Technical Papers
9. Kim ET, Lee JY, Kim YT (2009) Investigation of electrical characteristics of the $In_3Sb_1Te_2$ ternary alloy for application in phase-change memory. Phys Status Solidi (RRL) 103–105
10. Ahn JK et al (2010) Metalorganic chemical vapor deposition of non-GST chalcogenide materials for phase change memory applications. J Mater Chem 1751–1754
11. Sebastian A, Gallo ML, Krebs D (2014) Crystal growth within a phase change memory cell. Nat Commun 4314
12. Junsangsri P, Han J, Lombardi F (2017) Design and comparative evaluation of a PCM-based CAM (Content Addressable Memory) cell. IEEE Trans Nanotechnol 359–363
13. Vatajelu EI, Pouyan P, Hamdioui S (2018) State of the art and challenges for test and reliability of emerging nonvolatile resistive memories. Int J Circuit Theory Appl 46(1):4–28
14. Mohammad MG (2011) Fault model and test procedure for phase change memory. IET Comput Digital Tech 5(4):263–270
15. Pirovano A, Redaelli A, Pellizzer F, Ottogalli F, Tosi M, Ielmini D, Lacaita AL, Bez R (2004) Reliability study of phase-change nonvolatile memories. IEEE Trans Device Mater Reliab 4(3):422–427
16. El-Hassan NH, Nandha Kumar T, Almurib HAF (2016) Implementation of time-aware sensing technique for multilevel phase change memory cell. Microelectron J 56:74–80
17. Lai S (2003) Current status of the phase change memory and its future. In: IEEE International Electron Devices Meeting. IEDM'03 Technical Digest. IEEE, pp 10–11
18. Lin L-C, Sheu S-S, Chiang P-C (2007) Sensing circuit of a phase change memory and sensing method thereof. US 11/968,041, Dec 2007
19. Nair PJ, Chou C, Rajendran B, Qureshi MK (2015) Reducing read latency of phase change memory via early read and turbo read. In: 2015 IEEE 21st International Symposium on High Performance Computer Architecture (HPCA), Burlingame, CA, pp 309–319
20. Berger J (1961) A note on error detection codes for asymmetric channels. Inf Control 4(1):68–73

21. Lavizzari S, Ielmini D, Sharma D, Lacaita A (2008) Transient effects of delay, switching and recovery in phase change memory (pcm) devices. In: IEEE International Electron Devices Meeting, 2008. IEDM 2008
22. Jurasek RA, Willey AD (2014) Multilevel differential sensing in phase change memory. US 14/223,199, Mar 2014
23. Happ TD, Lung HL, Nirschl T (2007) Current compliant sensing architecture for multilevel phase change memory. US 11/620,432, Jan 2007
24. Li J et al (2011) A novel reconfigurable sensing scheme for variable level storage in phase change memory. In: 2011 3rd IEEE International Memory Workshop (IMW), Monterey, CA, pp 1–4
25. Lin W-P, Sheu S-S, Chiang P-C (2003) Verification circuits and methods for phase change memory array. US 13/934,954, July 2003
26. Balasubramanian M (2009) Phase change memory: array development and sensing circuits using delta-sigma modulation. Thesis, Boise State University, Summer 2009
27. Ande HK, Busa P, Balasubramanian M, Campbell KA, Baker RJ (2008) A new approach to the design, fabrication, and testing of chalcogenide-based phase-change nonvolatile memory. In: Proceedings of the 51st Midwest Symposium on Circuits and Systems, 10–13 August 2008, pp 570–573
28. Campbell KA, Anderson CM (2007) Phase-change memory devices with stacked Ge-chalcogenide/Sn-chalcogenide layers. Microelectron J 38:52–59
29. Markus J, Deval P, Quiquempoix V, Silva J, Temes GC (2006) Incremental delta-sigma structures for DC measurement: an overview. In: IEEE Custom Integrated Circuit Conferences, 10–13 September 2006, pp 41–48
30. Oliver J, Lehne M, Vummidi K, Bell S (2008) Raman low power CMOS sigmadelta readout circuit for heterogeneously integrated chemoresistive micro-/nanosensor arrays. In: IEEE International Circuits and Systems Symposium (ISCAS), May 2008, pp 2098–2101
31. Baker RJ (2008) CMOS: circuit design, layout and simulation, 2nd edn. Wiley-IEEE, pp 483–504
32. Baker RJ (2006) Resistive memory element sensing using averaging. U.S. Patent Number 7,133,307, 7 Nov 2006
33. Terada Y et al (1989) 120 ns 128Kx8-bit/64Kx16-bit CMOS EEPROM's. IEEE J Solid-States Circuits 24(5):1224–1249
34. Kuo C et al (1992) A 512-Kb flash EEPROM embedded in a 32-b microcontroller. IEEE J Solid-States Circuits 27(4):574–582
35. Papaix C, Daga JM (2002) A new single ended sense amplifier for low voltage embedded EEPROM non volatile memories. In: Proceedings of the 2002 IEEE International Workshop on Memory Technology, Design and Testing (MTDT2002), pp 149–153
36. Sinha M et al (2003) High-performance and low-voltage sense-amplifier techniques for sub-90 nm SRAM. In: Proceedings of the International Systems-on-Chip Conference (SOC 03), pp 113–116
37. Ielmini D, Sharma D, Lavizzari S, Lacaita AL (2009) Reliability impact of chalcogenide-structure relaxation in phase-change memory (PCM) cells-Part I: experimental study. Proc IEEE Trans Electron Devices 56:1070–1077. http://dx.doi.org/10.1109/TED.2009.2016397
38. Li J, Luan B, Lam C (2012) Resistance drifting phase change memory. In: Proceedings of the IEEE International Reliability Physics Symposium, pp 1–6. https://doi.org/10.1109/irps.2012.6241871

Summary

Emerging NVMs can open several avenues for next generation computing. For example, they can replace SRAM based Last Level Caches (LLC) and DRAM based main memory in High-Performance Computing (HPC). For energy harvest-ed Internet-of-Things (IoT) emerging NVMs can replace energy-intensive Flash memory to maintain system state between powering cycles. Some of the emerging NVMs also offer storage of more than 1 bit to achieve high density. One of central challenges to widespread adoption of these memory technologies is the reliable, fast and energy-efficient sensing. This book provides a treatment of this topic for several popular classes of emerging NVMs. The sensing techniques for spintronic, phase change, resistive and ferroelectric memories are described in detail. Various complex trade-offs are highlighted with respect to the sensing techniques.

© Springer International Publishing AG, part of Springer Nature 2019   103
S. Ghosh (ed.), *Sensing of Non-Volatile Memory Demystified*,
https://doi.org/10.1007/978-3-319-97347-0

Index

© Springer International Publishing AG, part of Springer Nature 2019   105
S. Ghosh (ed.), *Sensing of Non-Volatile Memory Demystified*,
https://doi.org/10.1007/978-3-319-97347-0

Printed in the United States
By Bookmasters